HOME AND APARTMENT SECURITY

HOME AND APARTMENT SECURITY

Al Griffin

Henry Regnery Company · Chicago

Library of Congress Cataloging in Publication Data

Griffin, Al.
Home and apartment security.

Includes index.
1. Dwellings—Security measures. 2. Apartment houses—Security measures. I. Title.
TH9745.D85G75 690 74-31379
ISBN 0-8092-8322-0
ISBN 0-8092-8321-2 pbk.

Published by Henry Regnery Company
180 North Michigan Avenue, Chicago, Illinois 60601
Manufactured in the United States of America
Library of Congress Catalog Card Number: 74-31379
International Standard Book Number: 0-8092-8322-0 (cloth)
0-8092-8321-2 (paper)

Published simultaneously in Canada by
Fitzhenry & Whiteside Limited
150 Lesmill Road
Don Mills, Ontario M3B 2T5
Canada

Dedicated to
Chief E. N. Davis,
Los Angeles Police Department

Contents

Introduction

When Willie Sutton was once asked why he robbed banks, he said, "Because that's where the money is." Alas, that was in the "good" old days, when banks were the only place worth knocking over. Now, the police B&E (breaking and entering) records are mushrooming to epidemic proportions, with the number of burglaries approaching three million a year, according to the FBI.

Thievery used to be a desperate measure used only as a last resort, but today the rip-off is often "a means of communication" in social protest. The only householder who is reasonably safe is the person who takes steps to protect his property. No one can prevent a professional burglar from getting into his house. Fortunately, there are very, very few true professionals active, and they pick their marks with great care. However, there is plenty the homeowner can do to deter the punks and junkies who infest the night.

Security is indeed for sale, in the form of numerous security systems on the market. One trouble with the "security industry," though, is that many manufacturers and distributors of systems that purport to protect the buyer are exploiting the panic atmosphere by ripping off the public themselves, peddling worthless devices that can be worse than useless if they give the homeowner a false sense of security. The sales tactics some of them use, and the prices they charge (especially a good many of the suede-shoe salesmen operating on a high-pressure, door-to-door basis), would give pause to an "honest" burglar.

"Security systems" can be anything from a twenty-nine cent light bulb for the front porch to an elaborate system of laser beam sensors monitored by closed-circuit TV scanners costing thousands of dollars. It all depends on what you can afford to spend (or what you can't afford not to spend) for protection. Remember—for the average urban household that has *no* security precautions, the odds are 1–4 that it will be burglarized within the next twelve months.

All this is not to say that one should run scared. But there's no point in not doing something about the ever-present threat, either. The purpose of this book, then, is to let the householder know what's going on in the security industry, and what it is doing to him as well as what it's doing for him.

The book was written with the co-operation of police departments in twenty-six cities from coast to coast, and much of it is based on interviews with victims of burglaries and with a number of fellows in various attitudes of contrition in San Quentin, Leavenworth, Joliet, and Sing Sing (a couple of whom have served as consultants to equipment manufacturers).

A number of punks were interviewed in municipal and county jails, too, but little sense could be made of what they said. Or do. But they do it.

Here's how to stop 'em—at least on *your* property.

Al Griffin

1

Primary Door Locks

"I work new housing projects exclusively because the builders put in the cheapest locks they can get. When the door has a spring latch, I can open the average locked door in three to four seconds"—Convict at Joliet State Penitentiary, Joliet, Illinois.

Spring Latches

According to the FBI, over 75 percent of all burglaries involve intrusion through the front or back door. This statistic can largely be traced to an invention called the spring latch, which has a closure bolt cut on a diagonal to allow anyone going through the door to shut the door and have it lock automatically. Inasmuch as most residential entry doors shut toward the outside, the only thing anybody needs in order to slide the bolt (tongue) back in its channel is something to press against the tongue (when the door is shut, the bevelled-cut bolt is pressed back by the in-frame strike plate). An intruder only need use something thin enough to slip between the door and the frame, such as a plastic credit card, to press against the slant of the bolt face to make it back out of the striker—and the door is open.

The spring latch gives the householder about as much protection as a hook-and-eye closure from the dime store—the kind usually used on the inside of screen doors—less, if the hook fits snugly into the eye. Yet the spring latch is a great convenience item because the householder does not need to use a key to lock his door when he leaves the house.

Because he uses a key to get back in, the householder usually assumes that his door is safely locked while he is away.

Well, a spring lock will keep out little kids who could otherwise open the door by just turning the doorknob. But *all* of the burglars who committed the 2.5 million burglaries in the United States last year know how to open a door locked with a spring latch.

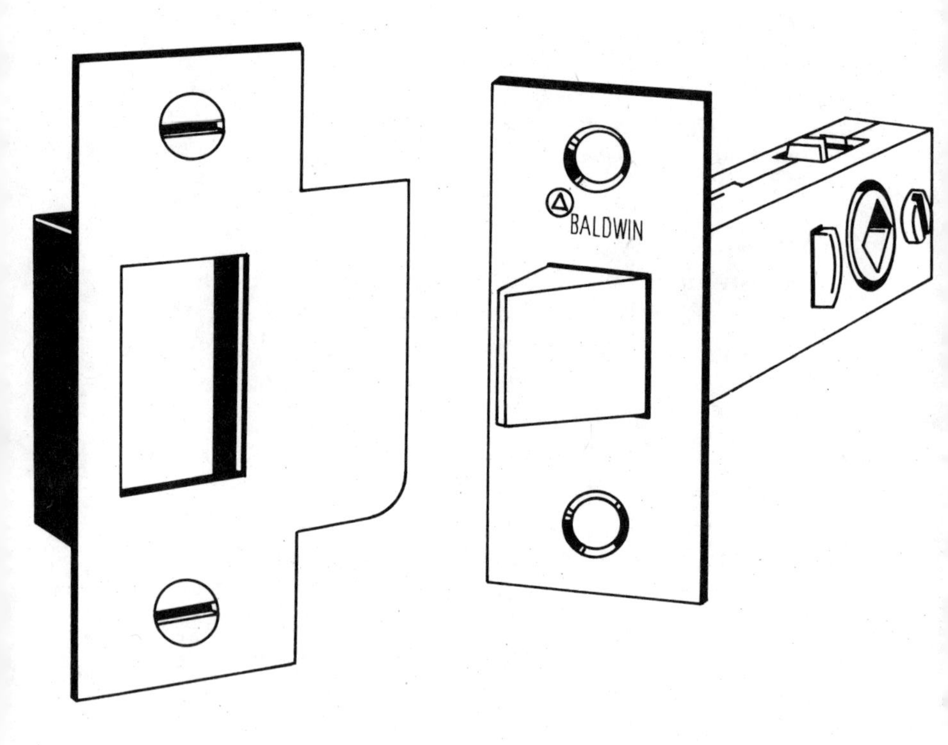

The spring latch is a great convenience for people in a hurry to shut and "lock their doors" without using a key, but is also relatively easy to *open* without a key by sliding back the latch with the pressure from a plastic credit card against the beveled face of the bolt.

Even the strip of wood the door bangs against when it closes affords little protection. Most doors are hung so loosely that there's room enough for the plastic card to bend around the buffer strip. If not, the wood strip usually is fastened to the 2"x4" door frame with small nails: thus it can easily be pried out away from the frame, sometimes even with a pocket knife, to provide enough room for the plastic card to go straight in at the spring latch.

Precautions

A tight door, with not enough room for a plastic card to bend around in, is the first requirement. But it can be completely "card-proofed" by screwing a flat protective metal plate into the frame against the inside of the buffer strip, to ward off any attempt to get at the spring bolt from the outside.

Another device to combat carding is a deadbolt.

Deadbolts

The original mortise locks made as builders' hardware usually incorporate a deadbolt, i.e., a flat-sided rectangular bolt that must be turned into a strike plate with a key. You can press the flat side of a deadbolt forever without making it move. But the overwhelming majority of homeowners, even the people who throw the deadbolt from the inside with the "night latch" knob, simply do not bother to use the key to throw the deadbolt when leaving the house ("Why should I, when the door locks automatically when I shut it?"), and a dismaying number of people do not even know that they *can* lock the door with the deadbolt from outside.

Although a flat-sided deadbolt can be cut with a hacksaw, very few burglars are willing to spend the necessary time. Thus, the round, free-turning deadbolt, which *cannot* be sawed through and is used by banks and other top-security operations, is seldom needed on a

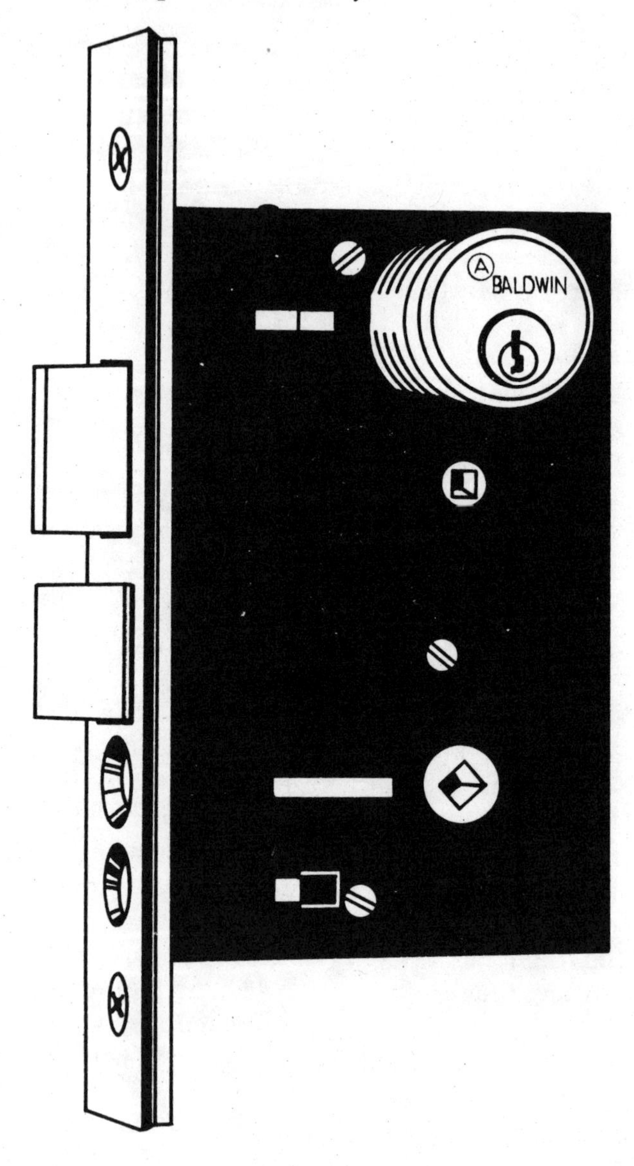

The best residential door locks combine a spring latch (bottom bolt) with a key-operated deadbolt (at the top). The flat-sided deadbolt cannot be carded for illicit entry.

residence. House burglars are lazy, sometimes scared, and often amateurs, and if they encounter any unexpected resistance they will go elsewhere for a hit.

Deadbolts may make nearly foolproof locks. But you do not need to unlock the deadlock to open the average door. That's what a burglar's jimmy is for. A jimmy is simply a pry bar that is used to push the framing wood away from the door far enough to disengage the bolt from the strike plate. Door frames are usually built of 2"x4" (actually only 1-3/4" x 3-3/4") lumber. A piece of wood that thin and up to seven feet long will have enough spring in the middle, where the strike plate is mounted, to be bent away from the door with surprisingly little pressure.

Most deadbolts are so short that they barely go into the strike plate's cavity when the doors are locked, some just barely catch. A trained fireman, using the edge of his ax as a jimmy, can open the average deadbolted door so fast that he is inside almost without breaking stride. And if a fireman can learn that technique, so can a burglar. (This writer did, spending an interesting afternoon jimmying doors with everything from a crowbar down to and including a common screwdriver.)

The householder can protect himself against jimmy artists simply by installing a deadbolt with a full one-inch throw. A bolt that long may require the installer's temporarily taking off the strike plate to hollow out some more of the wood to make room for the thrown bolt. But the cost and effort will be rewarded. A one-inch throw effectively jimmy-proofs almost any door because the frame can't be sprung a full inch.

Installing an auxiliary deadbolt with a one-inch throw will cost from $10 to $50, depending on the quality of the lock mechanism, but an auxiliary deadbolt requires the use of an extra key and is therefore somewhat of a nuisance.

Most deadbolts are operated with a key from the outside and with a knob on the inside. Doors with windows in them, or with thin wood panels a burglar can break through so he can reach in and turn the knob, should be protected with auxiliary locks having double cylinders, which are key-operated from the inside as well as from the outside.

However, internal key-operated locks, while giving top security, are dangerous; locking the double-cylinder lock not only locks everybody out, it also locks everybody in. In case of fire, or any other emergency when people have to get out fast, nobody can get out unless they have a key. People with double-cylinder locks should keep a spare key close to the door (out of arm's reach from the door) for emergencies, but turning a knob is a lot faster.

Use of an interlocking ring-and-bar lock with a vertical bolt also makes a door impossible to jimmy. Instead of having a slotted receiver in a strike plate for the conventional latch or bolt, the dual receiver accepts a vertical bolt that drops down through two heavy rings mounted against the inside of the door frame. The receiving rings are part of a right-angled metal plate mounted with as many as seven heavy-duty screws. The part with the vertical bolt, mounted on the door, is thus locked *onto* the frame, rather than merely into it. A door with this kind of a lock cannot be pried away from the frame, and its only disadvantage is its ugliness, regardless of which manufacturer makes it.

This type of jimmy-proof deadbolt has been made in such quantities by the Segal Lock Company that the name "Segal" is now used almost as a generic term. Cost is usually only $10 to $15, including the best. The lock made by Eaton, Yale and Towne, even has an internal shutter guard to prevent forced entry even if the outside cylinder is pulled (see below).

Keyhole Door Knobs

Installing a good auxiliary lock is particularly advantageous for the owners of many newer homes where the

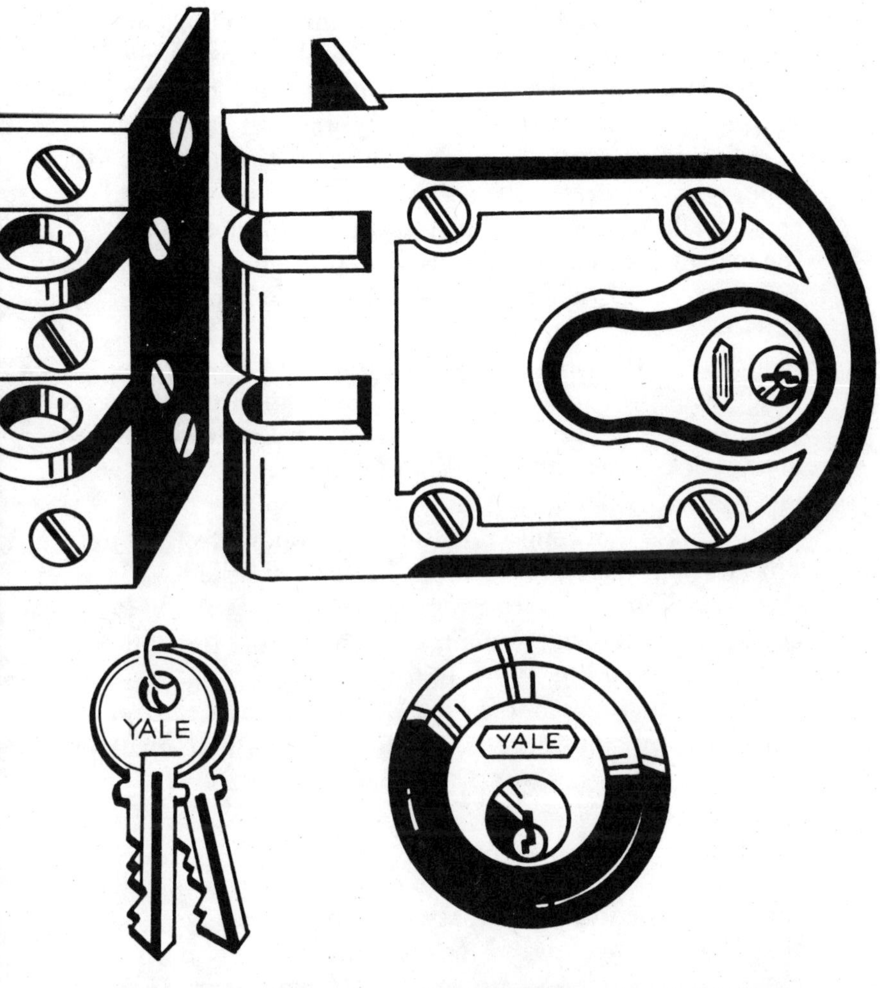

Jimmy-proof ring-and-bar lock has a vertical bolt that is impossible to pry away from the frame.

doors are built with the keyhole in the center of the doorknob as a cost-saving device. Some homeowners also like keyholed door knobs because the keyhole is easier to find in the dark.

All such locks are particularly valueless. A burglar with a heavy hammer, for example, can break off the whole door knob with one blow. He thus exposes the locking mechanism inside, which is then easy to pull out entirely.

Intruders equipped with a burglar's tool called a lock rape, which operates like a garage mechanic's valve puller, can usually pull out the deadbolt lock cylinder. However, the lock rape can be foiled with a security lock plate, which fits over the lock, leaving room only for the key. The case-hardened plate is bolted to the door with carriage bolts, and a typical lock plate—Fruh's Ace Lock Plate, for example—can be installed in a matter of minutes. All that is required is that two 5/16" holes be drilled through the door. Lock plates cost only $3 or $4 and are always a good investment.

Another tool commonly used by burglars is an ordinary pipe wrench, which is used on doors where the lock keeps the door knob from turning. If the pipe wrench is big enough, it will turn the knob anyhow, simply breaking the internal mechanism by brute force.

A wrenchman is, of course, helpless when confronted with a keyhole-lock door that is protected with an auxiliary deadbolt.

Warded Locks

At one time *all* homes were equipped with deadbolts that had to be locked with a key from the outside, because that was the only kind of lock in existence, and millions of older homes still are protected only with warded locks, particularly on kitchen and basement doors. Unfortunately, all these old locks have warded lock mechanisms, and a warded lock, which operates with a shaft-and-tang key, is the easiest of all locks to pick (unless you count the type of lock used on a high school kid's diary, which doesn't even have wards inside and has been picked by countless thousands of nosy little siblings armed only with a hairpin).

In a warded lock, the key tang engages the bolt directly and slides it back into the door as the key is turned. So that just any tanged key won't work, the mechanism incorporates metal projections to ward off contact with the bolt by alien keys. The key must be cut so as to allow it to pass over the wards when it is turned.

The chief trouble with warded locks is the fact that anyone can buy skeleton keys in any dime store and then can open "private" warded locks in literally millions of homes. A skeleton key works because the tang is so thin that it simply evades the wards.

Although the common T-shaped skeleton key will open most warded locks, some manufacturers of warded locks make their locks as different from each other as possible, positioning the wards differently in various locks. But a burglar equipped with a full set of skeleton keys—five or six different types at the most—can open virtually any warded lock by trying one key after the other.

Picking the lock with a strong metal strip about 3/16" wide is only a little more sophisticated than using a skeleton key to open a warded lock; all the user has to do is bypass the wards to flip back the bolt.

A warded lock should be considered as a privacy lock at best, never as a security lock. However, when warded locks are used properly, in applications requiring minimal security, they actually have some advantages over cylinder or tumbler locks. They not only cost a great deal less money, but they also are more impervious to freezing and clogging by dirt than are the more sophisticated mechanisms. Finally, the mechanism in a warded lock is so simple that it practically never gets out of order.

Cylinder Locks

Any cylinder lock is much harder to pick than even the best warded lock. This is because (1) the key turns only the cylinder and does not make direct contact with the bolt

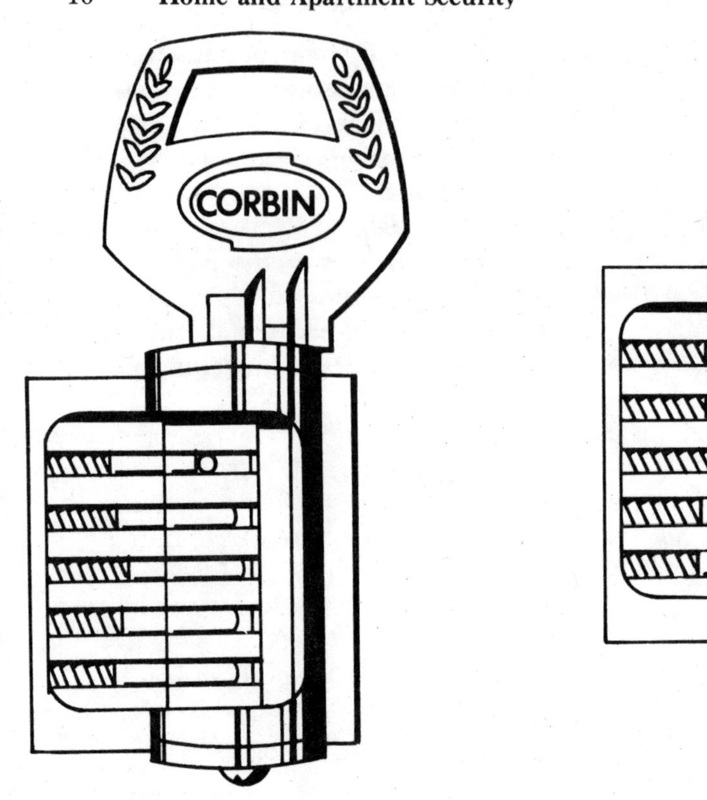

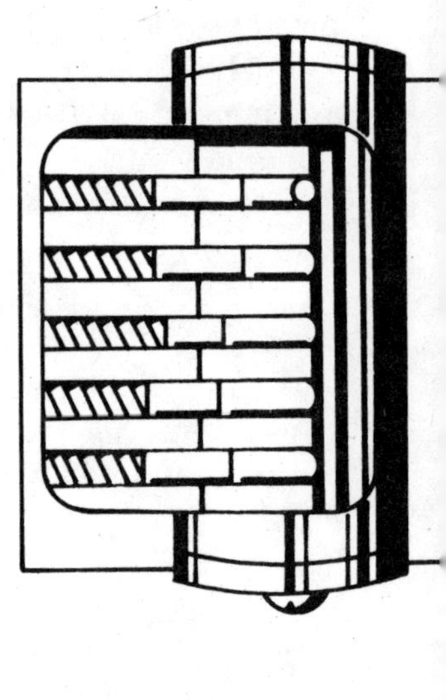

Cylinder lock with a pin tumbler action has a plug that is locked into the shell by the pins protruding into it. The proper key simply lines up the pins evenly at the top of the plug so that it can rotate as the key is turned.

itself, and (2) the burglar has to pick not just one mechanism but has to manipulate from three to seven pins (tumblers) one at a time. One of television's many disservices has been to make lock-picking look easier than it really is. In fact, only somebody such as Alexander Mundy, in "It Takes a Thief," can insert that 3/16" metal strip into a keyhole and open a cylinder lock with a single click in the right place. Picking a cylinder lock requires work on each individual pin, since every pin keeps the cylinder from turning.

In a cylinder lock, the keyhole itself is slotted (warded) to prevent the insertion of just any single-bitted key, with slots cut into the flat sides of the keys to allow the proper key to go into the keyhole. Notches cut into the edge of the key (bitting) then activate interior discs or pins, and if the pins fall (tumble) into the notches to the proper depth, they line up evenly across the edge of the cylinder, allowing it to turn.

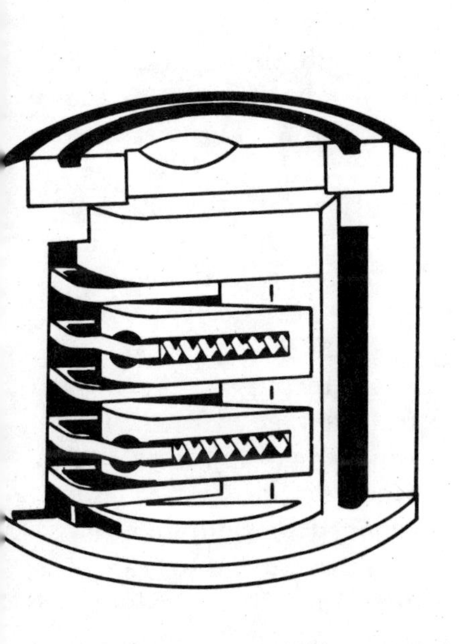

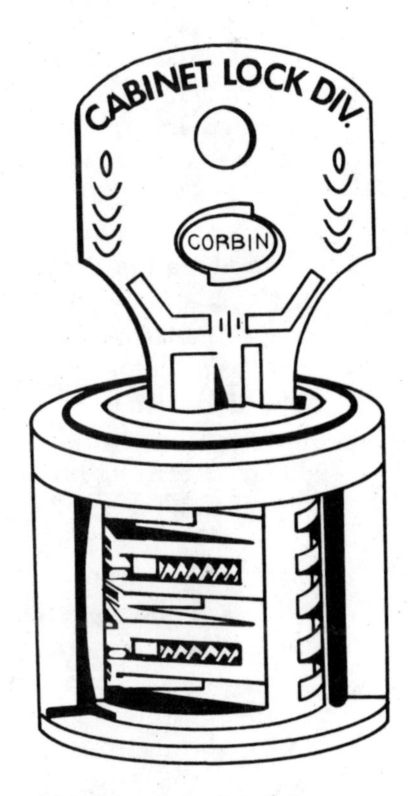

Low-cost disc cylinder lock has a plug from which discs protrude into the shell to keep the plug from turning until the proper key is inserted to retract the discs into the plug.

Lower-cost disc cylinder locks are relatively easy to pick, even with that ubiquitous thin metal strip, because each of the three to five discs can simply be turned until they are all lined up properly. But in a pin cylinder lock, each of the three to seven pins (most usually, five) keeps the cylinder from turning with its own pressure spring. Each is a two-piece pin, and the separation between the two pieces must be lined up with the edge of the cylinder before the cylinder can move. The lower part of each pin varies in length from that of the other pins in the lock.

Picking Cylinder Locks

Picking a pin-tumbler lock requires a burglar's tool known in the trade as a Yale pick (the pin cylinder lock was invented by Yale, although the patents have long since expired and pin-cylinder locks are now made by any number of other manufacturers as well). A Yale pick is a flat piece of metal small enough to fit into the keyhole used in combination with a stiff piece of wire bent up at the business end to a 90° angle. Using the metal insert as a wrench to keep pressure on the cylinder, the thief tries one pin after another with the pick until he finds the pin that first binds. He then releases wrench pressure enough to let the upturned wire push up the pin, and with delicate enough pressures he can feel when the bottom part of the pin comes flush with the top edge of the cylinder. He repeats the process with each pin until their separations are all lined up properly to allow the cylinder to turn. The turning cylinder operates a cam projecting from the back of the cylinder to retract the bolt out of the strike plate.

In an exceptionally well-machined lock mechanism, the thief cannot depend on loose manufacturing tolerances to let him feel the proper lift, in relation to the cylinder edge, for each pin. In that case he has to raise each pin slightly, one at a time, over and over again, consuming valuable time, until the top parts of the pins come up out of the cylinder all at the same time to allow rotation of the cylinder.

Even after a lot of practice, this author found it difficult to pick most ordinary pin-cylinder locks in much less than half an hour. Best time: 4½ minutes.

Some burglars claim that they can do it in seconds, even though nine out of ten burglars know nothing about locks and depend on public negligence rather than skill. Of the sixty convicted burglars (or awaiting trial) interviewed during the author's research for this book, not one knew anything about picking pin-cylinder locks. However, remember that these sixty burglars were losers; they got caught. The experts are caught less often.

According to the FBI, the average burglar is 18½ years old, he has very little education, and he spends less than seven minutes on the entire typical job from the time he approaches the house to the time he flees. But it should be remembered that FBI figures are based on the records of *arrested* burglars. The FBI knows nothing whatever about the ones who have never been apprehended. And for the burglar who is careful to avoid confrontation, burglary is one of the safest of crimes to commit because nobody can identify him. It can be said unequivocally that most burglars are *not* in jail. The FBI says that less than 20 percent of burglaries are ever solved, and these are the crimes that are most likely to involve lock picking.

Pick-Resistant Locks

Thwarting the burglar who does know how to pick a lock is most easily accomplished by using more pins in the cylinder. The more pins there are, the longer the lock takes to pick. Another way some manufacturers make picking more difficult is to use mushroom heads on the pins. These heads catch in the corners of their holes when pushed up with a pick instead of with a key. Time is very important. Professional burglars do not like to spend much time trying

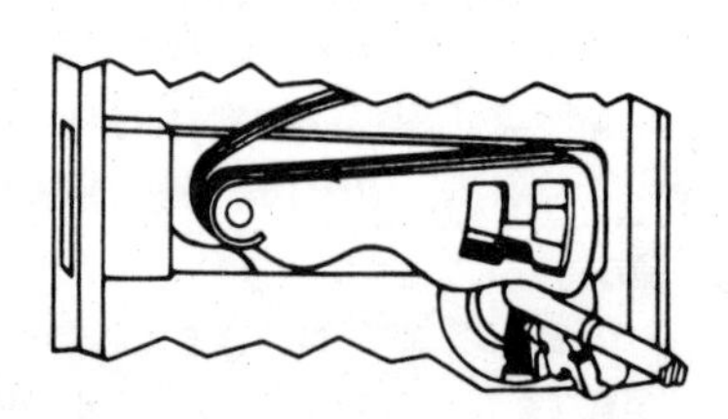

Bolt unlocked. To operate, the lever tumblers must be aligned so as to provide a "gate" through which the "fence pin" (attached to the bolt) may pass.

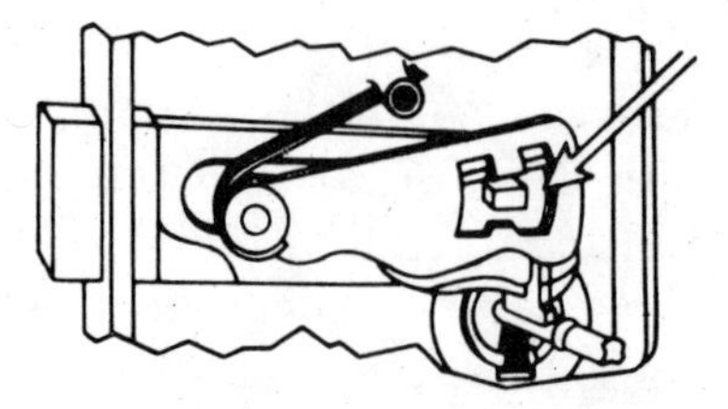

The key has aligned the lever tumblers so that the "fence pin" is passing through the "gate," and the bolt is thus moved.

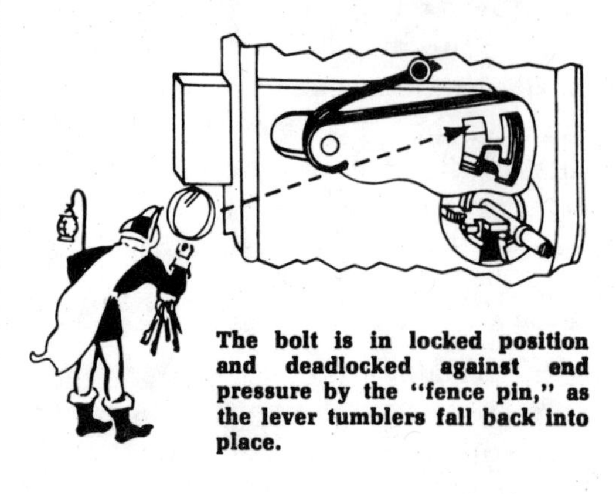

The bolt is in locked position and deadlocked against end pressure by the "fence pin," as the lever tumblers fall back into place.

Lever tumbler locks have so many moving parts, and can get out of order so easily, that they are seldom used on anything but interior doors where they are not subject to varying weather conditions.

to get in; it is too easy to be seen by a passerby.

Several special pick-resistant locks on the market will discourage all but the most determined burglars. None of these locks is used as builder's hardware because they cost too much, but they can be installed by a locksmith to replace original cylinder mechanisms.

The Duo maximum security lock, made by the Illinois Lock Company, Wheeling, Illinois, costs from $17.95 to $21.95, but it is exceptionally hard to pick because it has a full complement of fourteen pins—two sets of five primary pins plus a secondary set of four. The design provides for mechanical movement of the tumblers without the aid of springs. Its triple-bitted key has not just the ordinary single set of notches; it has three—one set on each side of the key and one on a "terrace." Instead of being made of brass, as

Key for a Duo maximum security lock has a triple-bitted design . . . one set of notches on each side of the key and a separate set on the "terrace." These notches activate fourteen separate pins in the lock, making it extremely difficult to pick. A Duo key cannot be duplicated on ordinary key-making machinery.

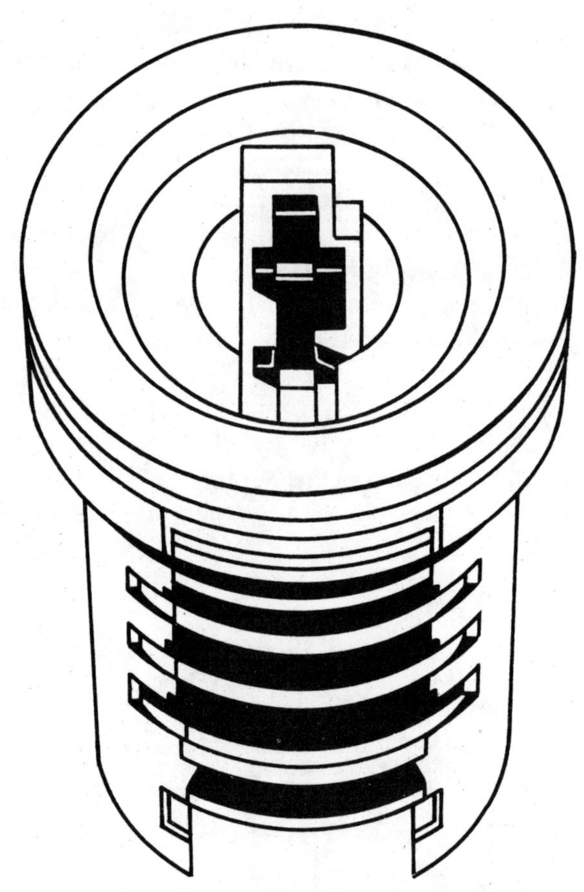

Duo lock presents formidable opening for key entry, and ordinary "master keys" will not work. Shell contains two sets of primary pins, five per set, and a secondaray set of four more pins as extra security.

are most ordinary locks, a Duo has a hardened steel keyhole system that will break any common drill or pulling device before it can penetrate. A Duo lock has 64,000 different possible key changes. All Duo keys are registered and cannot be duplicated on standard key-making machines; the key codes are completely restricted; and the homeowner can get extra keys only by going direct to the factory or to one of its authorized locksmiths.

Illinois Lock makes another lock not quite so formidable as the Duo, but difficult to pick nevertheless—the "C" model, which has two sets of four tumblers each, in opposite barrel spines. Its double-bitted key, also providing direct cam action without the aid of springs, can be duplicated only by locksmiths capable of cutting notches on both sides of the key.

Sargent Lock Company, New Haven, Connecticut, makes the Keso pick-resistant lock—with three complete sets of tumblers arranged around the keyhole radially. Picking these three sets of pins while keeping critical wrench pressures exact is so complicated a process that even an ordinarily good pickman usually will seek easier pickings elsewhere when confronted by a Keso. The unique hexagonal key is not notched, but its depressions are drilled to different depths in its six sides. Try getting that one duplicated at Woolworth's!

Chicago Lock Company is also strong on unconventional keys for its maximum security locks, one being the tubular key. The tubular key fits into a round, centerwarded keyhole; its unique shape alone is enough to make even a fairly skilled burglar decide not to invest his time on anything so unfamiliar. The pin tumblers are not lined up in a straight front-to-back row as in an ordinary pin tumbler lock, in which an expert can "rake" the pins. In the Chicago Ace Lock, the pins are set in a circle facing the front of the lock, and the working rim of the tubular key is cut in slots to various depths around its circumference to depress the seven pins accordingly.

Finally, the Chicago Ace key has a projection at the top of its tube to match a notch in the top of the keyhole, so that the homeowner can get the key straight in, with the proper cuts going direct to the proper pin tumblers.

These tubular keys, of course, cannot be duplicated

Sargent's Keso lock requires a hexagon-shaped key, with depressions drilled to different depths on its six sides, instead of having notches. The key operates three complete sets of tumblers arranged around the plug radially, making the lock virtually impossible to pick.

even by locksmiths unless they are authorized to do so by the factory.

Chicago Lock also makes a "hidden tumbler" lock, utilizing a seven-pronged tubular key for still another unconventional-looking keyhole.

There's no such thing as a pick-proof lock, of course, if an expert has enough time to work on it. But *the* toughest

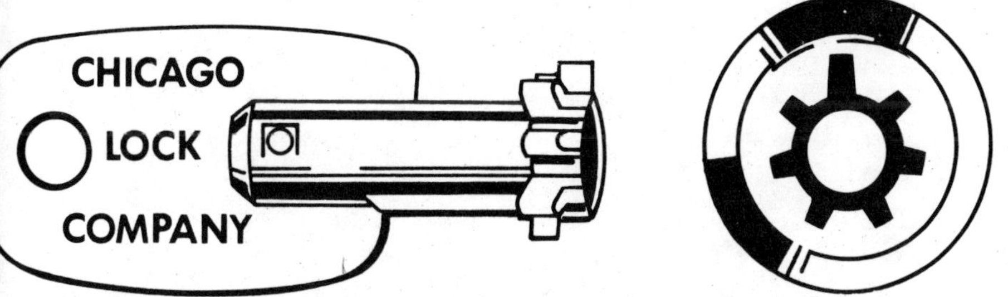

Tubular keys are widely used to lock or disarm burglar alarms. Pins locking the shell are set in a circle facing the front of the lock, which makes lock picking extremely difficult.

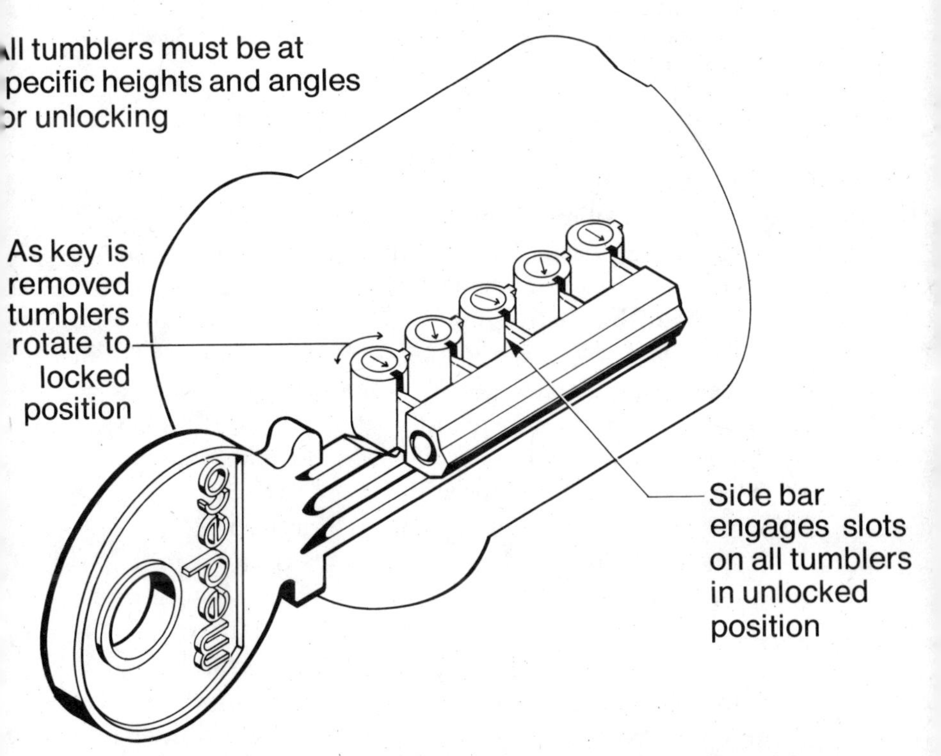

The Medeco lock, long known as *the* pick-proof lock, utilizes pin tumblers which not only must be at specific heights to allow the cylinder to turn but also must be turned individually at specific angles.

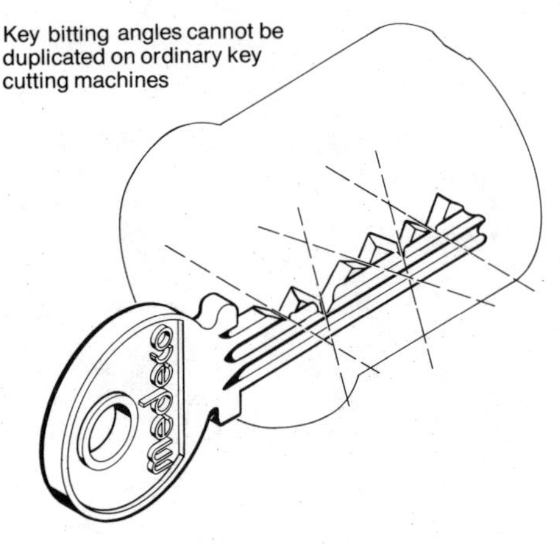

Notches on a Medeco key are bitted at individual angles as well as to various depths. Medeco keys cannot be duplicated on ordinary key-cutting machinery.

lock to get open without a key, by long odds, is the pin-cylinder mechanism made by Medeco Security Locks, Inc., Salem, Virginia. The inventor was so confident in the Medeco's superiority over anything else in the maximum security field that he originally offered a $10,000 prize to anybody who could pick the lock—amateur or professional. However, he overlooked putting a time limit on the attempt and soon had claimants left and right who said that they had indeed been able to pick the lock—in, like, four hours. Medeco's offer was quickly withdrawn, and the lock is no longer sold as pick-proof, but it is indeed pick-resistant.

The secret of the Medeco is in the five *rotating* pin tumblers. Each pin has to be turned correctly before it can be pushed up from the cylinder plug. Furthermore, the twisting pins are locked into place by an internal side bar, which prevents them from rotating. Several hundred times more key changes than in a conventional pin-cylinder lock are possible. Just for good measure, the critical components are made of case-hardened steel to prevent drilling.

The key for a Medeco lock is a real weirdo. At first glance it looks like any other key for a cylinder lock, but close examination reveals that each notch cut into the side of a Medeco key is cut at a different angle, a key bitting technique impossible for ordinary key-cutting machines. If a householder needs extra keys, he has to deal with the factory or a licensee. If all keys are lost, not even a licensed Medeco locksmith can make replacement keys, even by taking the lock apart. Such an unfortunate householder has to send a written request to Salem, where even his signature is registered, asking for keys made to his registered code.

Medeco locks are available only through authorized Medeco locksmiths and cost $25 to $30, depending on style. They are worth every penny to anybody who wants to replace ordinary cylinder locks for maximum security.

Close to the Medeco for pick resistance are the double-security pin cylinder locks that not only need the mechanical action of a key but that also keep the pins in locked position magnetically. The magnets can be released only by a specially coded key, which, again, cannot be duplicated in an ordinary key shop.

Key shops are also out of the picture for computer-type locks, which operate with plastic punch cards instead of with recognizable keys. Each "key" has an average of twelve pinholes that match up with electric eyes, magnetic fields, or electronically sensitive "receiver plates" to release the catch. Most such systems are in the $50 to $100 per door range.

There's even a lock mechanism on the market that has no keyhole (or card slot) of any kind; it works with push buttons. The householder simply pushes the right sequence of numbered buttons to retract the mechanically operated latch.

The push-button lock made by Preso-Matic Lock Company, Lyons, Illinois, is made in both spring-latch and deadlock styles and costs from $25 to $40 per door. The

Push-button locks, which operate without keys, work like a push-button telephone. When the homeowner uses a four-number combination, the lock provides 10,000 different possible combinations. Nervous homeowners (with good memories) can switch the ten-button lock to operate on a sequence of seven numbers, to increase the odds to some ten million to one.

"night latch" consists of a button on the inside, which makes the ten outside push buttons inoperative, even for somebody who does know the combination. The door is opened from inside by pressing a single "unlock" button. Locking the door from the outside involves pressing a reset bar.

Burglars facing the usual four-number combination are confronted with 10,000 possible combinations. We quit trying to hit the right combination after trying only 1,000 . . . which took over two hours. The ten-number push-button lock can also be switched over to operate on a sequence of seven numbers—when the odds against finding the proper sequence become about 10 million to 1.

Using a seven-number combination can have adverse consequences, however. Children and old people may have trouble remembering a number that long.

Owners of push-button locks sometimes use a ploy that foils would-be intruders even more than just the odds; they "hide" a slip of paper under the doormat or in the mailbox marked "Combination to push-button lock: 8471" or whatever—except that the number is a fake. Nobody trying that four-digit number will persist for very long.

Another dodge, which is especially useful when the house is occupied, is to wire up the doorbell to one of the buttons not used in the combination so that anybody monkeying around with the buttons will anounce his presence himself.

If the kids blab the combination to too many of their pals, or if for any other reason the householder wants to change the combination, he can do it himself in a matter of minutes (the Preso-Matic people say five minutes, but it took us twelve minutes). It's a matter of removing an inside cover plate and replacing the two brass "combination slides" with two new ones, which cost only $2.50 a pair.

This is admittedly a much simpler process than what a wary homeowner goes through when somebody in the family loses a house key. With a push-button lock, there are not only no keys to lose, but when "an extra key" is needed for somebody, the homeowner simply tells him (or her) the number.

Preso-Matic locks are heavy-duty keyless locks with one-inch deadbolts and case-hardened steel hardware. This is not always the case with many "space age" keyless lock systems. In addition, some are susceptible to adverse weather conditions and can be complicated enough to get out of order from time to time.

Anti-Intrusion Locks

People who are *really* afraid of burglars can invest in a police brace, such as that made by the Fox Police Lock Company, New York, New York. The police brace is actually an anti-invasion device for protection against anybody trying to smash the door in by brute force, and it works on the same principle as wedging the back of a chair under the door knob. Basically, it is a long steel bar. The bottom end fits into a metal socket recessed into the floor about two feet away from the door; the top end fits into a lock mechanism mounted above the door knob. The brace can be locked or unlocked into or out of place from outside.

Inasmuch as there is no tongue going into the door frame, there is no point in trying to jimmy a door protected by a Fox police brace. However, the whole works—lock and keys included—costs only $18.95, which means that the lock is a cheap one and not much of a deterrent to anybody who knows how to pick locks or pull cylinders. It is obviously designed for protection against "amateur" burglars and common hoodlums.

Fox's lock cylinder can, of course, be replaced with anything up to and including a Medeco cylinder. Even then, if the door has an old-fashioned "peek through" keyhole for a warded lock, a crook can insert a coat-hanger wire to shove the top of the bracing bar to the side and out

A police brace is an anti-invasion device, making the door hard to kick in. It can, however, be locked and unlocked from the outside. Not very aesthetic, but very effective.

of its security slot. Even if the homeowner fills up the old keyhole with concrete, a burglar could achieve the same result by drilling a small hole in the door.

Fox also makes a $65 double-bolt lock, with the lock mounted in the door's center from which heavy bolts extend from both sides. It is most usually used on the back doors of stores and in factories, with the two bolts penetrating up to 2½ inches into the door frame on both sides—very effective, even if not very attractive.

Many residential doors, especially back doors installed by do-it-yourself homeowners who have remodeled back porches, open toward the outside. So do the doors to many apartments in, say, Victorian mansions that have been cut up into multi-unit dwellings. In all such cases, the best locks in the world are meaningless if the door hinges are on the outside. The only thing a burglar needs to do is pull the hinge pins up and out and then pull the door open "from the back."

The householder can counteract this possibility by drilling a hole through the surface of a hinge from the inside (open the door and do the work between the open door and the frame) and well into the hinge pin and by fitting in an arresting pin that will keep the hinge pin from being pulled out.

When the door is shut, there is no way for anybody to get at the arresting pin. The homeowner should leave enough of it sticking out of the hole to provide a grip for a pair of pliers, though, in case *he* needs to pull out the hinge pin.

Installations

No security hardware of any kind is any better than the way it is anchored to door and frame. One trouble with do-it-yourself homeowners who install their own locks is that they use anything at hand instead of getting the proper hardware. Many locks, for example, are attached from the

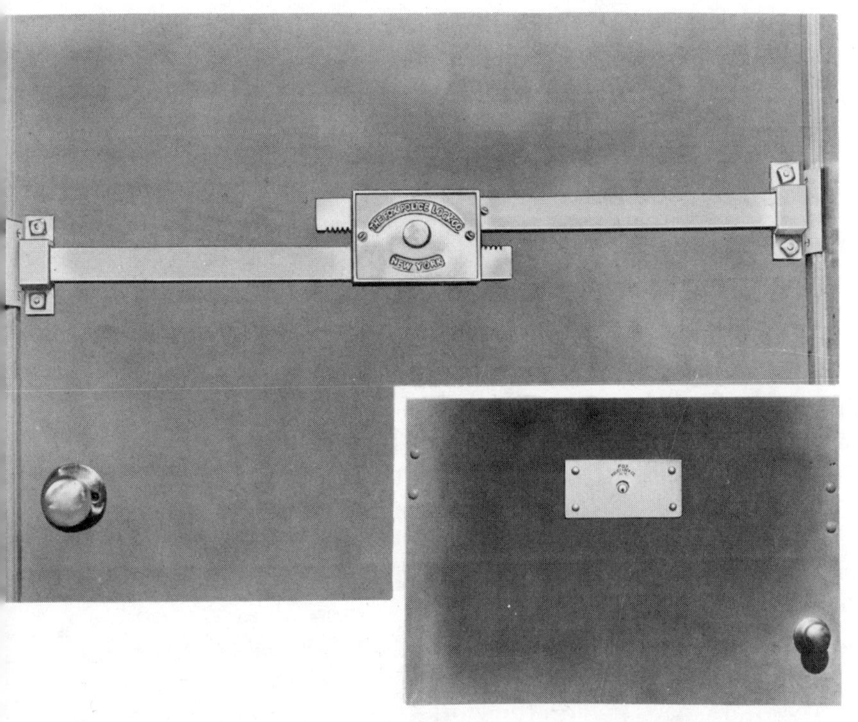

Fox Police Lock throws separate deadbolts into each side of the door for maximum security. The locking mechanism is vulnerable to picking, however.

outside with screws instead of carriage bolts, which means that anybody with a screwdriver can get in. Other do-it-yourselfers often use interior screws as short as 5/8 inch long; such screws can be jerked right out of ordinary pine door frames by just about anybody throwing his weight against the door.

Installation or replacement of a lock should be done by a locksmith who is either bonded or a member of the Associated Locksmiths of America or of a regional locksmiths' association. The burglary details of local police departments are often able to recommend one. Locksmithing is an honorable profession going back thousands of years, and a good locksmith values his work and his reputation. He can install security systems that no do-it-yourselfer or carpenter would know anything about.

Doors

The door itself should be of solid construction. Altogether too many residential doors are built with a solid framework but with thin wooden panels in between, often only 3/8" thick. Some of the panels, especially at the corners where they meet the door's framework, can be smashed in with a gloved fist in a single hard blow. In seconds a small hole can be drilled to allow for the use of a compass saw. It takes very little extra time to cut a hole big enough for the otherwise unskilled burglar to get his hand in—at the inside knob—and open the door.

The flush doors used in most apartment buildings are ridiculously simple to cut through, consisting as they do of a hollow wooden frame covered with thin sheets of plywood inside and out. When they are made of metal, as required by fire regulations in many cities, the sheet metal is often a mere 1/64 inch thick; it is usually soft metal besides. This stuff can be cut with nothing more than a sharp hunting knife; a sharp blow for penetration and then muscled slicing are all that's required. Drilling a hole big enough to accommodate a hacksaw blade doesn't take much longer.

Some people fill up their hollow doors with iron plates or even reinforced concrete near the lock, but this can result in a door so heavy that its weight can pull flimsier hinge screws right out of the frame in day-to-day use. Most apartment owners frown upon any such measures, and the tenant who tries to make his door burglar-resistant in this way will often find that he has broken his lease.

"Right of Entry" Keys

Building superintendents do not like to have their

tenants put auxiliary locks on their doors, on the premise that the management has the right to enter any apartment in case of emergency (which is why the super usually has a master key or set of duplicate keys that will open any apartment in the building). If the super demands a duplicate key to the auxiliary lock, the tenant may be well advised to refuse. Sure, the super himself may be honest, but what if somebody burgles his flat and gets hold of the master key?

The normal lease says that the superintendent has "right of access." What this really means, legally, is that if he knocks on the door, the tenant has to let him in when he has a legitimate reason. It does *not* mean that the tenant has to be at the mercy of anybody with keys to the door.

Special leases may indeed state specifically that the super be allowed to hold a copy of *all* keys. There's even a reason for such a clause; for example, if a fire should start in an apartment while the tenant is out, it could endanger lives and property all over the building if the super cannot get the door open. But there's no reason to simply hand over a duplicate key that can possibly be used indiscriminately. The super's duplicate should be sealed in an envelope with the tenant's signature across the flap and all other closures, which should then be taped down with Scotch tape. He can get at the key in case there actually is an emergency. But if that envelope has ever been opened for any reason, the tenant will be able to tell at one glance.

The key is often the weakest link in a security system. Anybody moving into a new house or apartment should change the house locks, in case the former tenant (or builder) still has duplicate keys to the place. Keys should never have identification on them, in case they get lost. Why give a burglar your name and address along with the keys? Even the little automobile license plate replicas, often sent out in charity drives, should never be carried on a key chain, because finding out the owner's name and address from the state license bureau is too easy.

Finally, the best locks built are not much good if the door is left unlocked. An amazing number of "rattler" burglars just walk right in—there are over half a million illegal entries per year where no force whatever is needed to get in. Rattlers work mostly in the daytime when the tenants are at work, the kids are in school, and the house is empty, in a field where residential burglaries have increased by 74 percent during the latest five-year study made by the FBI.

In researching this book we took a number of locks apart to see how they work and also got schematics of various protective designs, but this book is no place for such information. Suffice it to say that if a commonplace writer, with no more criminal proclivities than most people, can find out in a few days how to get through 95 percent of all locked doors, so can anybody else who really wants to do so.

2

Secondary Locks

"One of the best hits is always the second time around. By six weeks or so after I have cleaned out a house, the people have replaced all the stuff I stole the first time with brand new merchandise, which always makes for a better score"—Convict, Sing Sing Prison, Ossining, New York.

Experienced burglars will go through a window only if they know what they are after inside, and whether it's worth taking a chance for. If a burglar is caught monkeying around at a door, he can come up with all kinds of plausible stories; but if he is seen crawling through a window, there is nothing much he can say.

But a burglar who does know what's inside the house, or an amateur who doesn't realize the chance he's taking, will prowl the entire premises looking for the easiest means of entry. This is often a window left open for ventilation, particularly in the summer—one reason why FBI statistics show that burglaries hit a high of about 10 percent above average in July (of course, this is also the season when more people are out of the house more often and for longer periods of time).

People who leave the storm windows on the year around (usually because they have air conditioning) have better window protection than the people who put up easy-to-slit screens. The old-fashioned storm window, which is fastened to the frame with six long screws (even though from the outside), affords particularly good protection. Taking out the screws takes time, as any cursing householder himself can testify every spring.

Screens

Inasmuch as a screen window or a screen door is so easy to cut through, there is no point in using good locks on such points of entry. But there's not much point in making it too easy for the burglar to take the screen out, either, such as using just hook-and-eye fasteners. These can even be an *inducement* for an amateur burglar who gets a kick out of playing cops-and-robbers. Any child can poke a hooked wire through the screen and pull the hook up and out of the eye "just like a big timer." A small angle iron screwed into the frame at each side of the window will at least deter the many intruders who are reluctant to cut their way in.

Even better, nail the screen to the inside framing. Thus, the intruder either must work at the nails with a claw hammer to get the screen out or will have to cut away enough screening to crawl through. If the householder leaves the nailheads sticking out enough for a claw grip, he can use the same nails and nail holes year after year.

Sash Windows

The window itself, in most homes, is a double-hung wooden-framed sash, having top and bottom halves that can be "locked" together with a thumb latch. Unless the fit between the top and bottom separation is unusually tight, this type of latch can usually be pushed open with a butter knife.

One of the most effective ways to secure a sash window is to fix it so that it can be opened only enough for ventilation, but not enough for anybody to crawl through.

The simplest way to do this is to drive a heavy nail into the track on each side above the lower window, with the nailheads sticking out of the frame about half an inch so that the window cannot be raised more than a few inches.

The same idea utilized to keep the top window from being lowered more than for ventilation is not as effective because the "stop" nails would have to be driven into the tracks on the outside where anybody with a claw hammer could pull them out. As an alternative for outside stops, use heavy screws and then cut off the heads so they can't be unscrewed. The householder himself, of course, will have difficulty getting such stops out of the wood if that ever becomes necessary.

If for any reason the bottom half of the window needs to be raised to its full height from time to time, you might consider drilling holes in the track big enough to accept removable steel dowel pins (heavy nails with the heads cut off). These dowel pins are not easily manipulated out of their sockets with a coat-hanger wire used by someone reaching through the ventilation opening, which is plenty big enough to get an arm through. Removable dowel stops can be inserted when the house is unattended, and then removed to allow the windows to be used normally when there are people in the house. Any emergency windows, such as those opening onto fire escapes, should have removable rather than permanently fixed pins.

An excellent way to lock a sash window in closed position is to drill a hole through the horizontal top part of the bottom window and on into the closed top half's horizontal member behind it. Then, by inserting a steel dowel pin a little smaller in diameter than the hole, you will keep the window halves from being either raised or lowered.

Though few people think of a toilet plunger as a burglar's tool, it is used effectively by a good many burglars to get through a locked window. The vacuum cup is pressed firmly against the glass near the lock, a glass cutter or sometimes just a diamond is then used to score the glass deeply around the plunger, and one good shove then breaks out a circle of glass big enough for the burglar to get his hand and forearm through. Adhesive tape is often used as an alternative, taped across the scored area after the glass cutter has been used. When the glass is knocked in, the adhesive tape will keep the glass from falling and making noise.

But such burglars can be stopped, even if they do go through the glass so they can reach in, open the catch, and try to raise the window. If the above-mentioned arresting pin is cut so that the end is flush with the wood, a burglar fumbling around with a gloved hand will not be able to find the hole, especially when he doesn't know what he's looking for.

If he does happen to find it, he will not be able to pull out a pin that is flush with the wood, even with needle-nosed pliers. When removal of the metal pin is necessary, the householder simply draws it out with a magnet—a tool very few burglars carry.

Safety Glass

Cutting a hole in the glass is as far as most burglars will go in attacking a window. A vandal could throw a brick through a window and smash out the entire pane, but that would make all kinds of racket—something most burglars try to avoid. For this reason, the laminated plate glass used in picture windows, which is not easy to cut through, is probably more burglar-resistant than most entrance doors.

For those homeowners who nevertheless are concerned about the possibility of somebody shattering an entire pane to crawl through, the ordinary window glass can be replaced with safety glass. This is most commonly the kind of "wire mesh" glass used in industry, with wire woven into the glass itself. Since it is ugly-looking stuff, its chief residential use is for basement or garage windows. Not only is it unattractive, but it is also expensive, costing more than twice as much as regular double-strength window glass, that is, about $2.50 per square foot.

However, industrial wire mesh glass is not the only safety glass available commercially. The Globe-Amerada Glass Company, Elk Grove Village, Illinois, makes a safety glass that is completely clear. Instead of making the glass smash-resistant with wire mesh, Amerada makes its Secur-Lite safety glass with a tough, high-tensile plastic inner layer that makes the glass incredibly resilient.

Secur-Lite was actually designed for commercial use, as in jewelry stores, since some 40 percent of all burglary attempts on stores are directed against display windows—according to Underwriters' Laboratories, which tested the stuff for resistance to smashing and battering. Globe-Amerada's safety glass is too expensive for homeowners to use on all windows (and there would be no point in using it only on one or two windows and leaving ordinary glass in the rest), costing around $6 per square foot, plus installation.

Casement Windows

Compared to double-hung sash windows, casement windows have limitations as well as advantages of their own. Casement windows are usually hinged at the sides and swing out from the house, with any screening inside the window. They are most often operated from the inside with a hand crank, which can only shut them, leaving them not locked with a catch of any kind. Such windows can be pried open quite easily, simply by applying a little leverage with a jimmy and springing the often fragile crank mechanism. To prevent such easy access, casement windows should always have an inside clamp, which the householder ought to use to lock the edges of the windows whenever he leaves the house, even if opening the screen to do so is a little inconvenient.

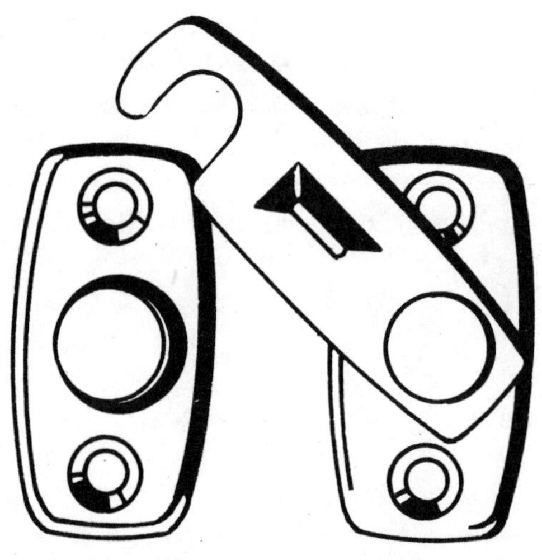

Decorator-type "cafe shutter bars" look better than hook-and-eye closures from the dime store, but the only extra security they offer is two screws at each side to resist forcible pressure.

For this purpose, a heavy hook-and-eye closure (for wooden casement windows) is better than nothing. Baldwin Hardware Manufacturing Corporation, Reading, Pennsylvania, makes some good-looking brass equivalents, which they call "cafe shutter bars." A bit better are tight-fitting neck bolts at top and bottom. These are surface bolts with an offset bolt that provides more space between the edge of the window frame and the entry hole in the receiver.

On a metal casement window, the best a homeowner can do is to put a pin of some kind into the catch. The metal is so thin that any commercial locks must necessarily be such lightweight types of closures, and any screws tapped into the framing so small, that they can be snapped open with one twist of even a small jimmy.

The best locking device for casement windows we have seen is one used by a Wisconsin homeowner on his double-opening metal casement windows. It is a steel bar as long as the full window width, running through a swiveling ring mounted with carriage bolts to the edge of the overlapping half of the double window. When the windows are open, the bar is held upright by a simple clip or slipped out of the ring altogether. When the windows are closed, the bar is turned through the ring horizontally, with the ends resting in curtain-rod clips at both sides of the frame.

These precautions are all the more important because, in most cases, going through the glass in casement windows, to reach in at a catch or clamp, is even easier than in sash windows. The burglar doesn't even need his toilet plunger or adhesive tape in many cases. Most casement windows, including many wooden ones but particularly metal ones, are comprised of six to nine small panes of glass. As in all windows, they are puttied into place with the putty on the outside (in the interests of keeping water leaks out). The burglar only need scrape out or chip away the putty around the pane nearest the lock, pull out the triangular metal glazier's points with a pair of pliers, and the glass will fall out in his hand.

On the other hand, multipaned windows do not allow a burglar to crawl through smashed (or removed) glass as in an ordinary double-hung sash, because the framings for the individual panes serve as bars to keep him out. If confronted with a keyed lock or other device that keeps the window closed, a persistent burglar would have to remove at least four panes of glass and then saw through the frames in at least four places. Few indeed are that persistent. Most will leave as soon as they find that they can't open the clamp or lock.

Basement Windows

Most small basement windows are wooden casement windows hinged at the top and opening inwards. The catches used as original equipment are the simplest kind.

Basement windows are deeply inset to keep out rainwater, and a burglar cannot get at the catches with a knife blade, as he can by shoving it up between the two halves of sash windows. But once he has cut a hole in the glass, he can usually open the catch with a flip of a finger.

The homeowner can put just about any kind of rim lock on such a window that he can put on a door (see Chapter 1). A surprisingly good substitute, judging by the experience of people who have had attempts made at their basement windows without the intruder's being able to get in, is a twenty-nine cent surface bolt (barrel bolt) mounted at each side of the window. Some barrel bolts, like the units made by Master Lock Company, can even be locked with a key. A burglar *could* open them easily . . . if he was willing to cut some more glass . . . and if he knew where they were. But the common reaction is to give up and try an easier method of entry.

Air Conditioners

Quite a few burglars specialize in getting into homes by lifting air conditioners out of windows. The first thing they cart away, of course, is the air conditioner itself, because there's always a lively market for used units in good condition. And the gaping window is then an open invitation to go in and browse.

Some air conditioners are mounted in such a way that they can be shoved *into* the room to provide for access. In other cases, the bottom half of the sash window, which was opened far enough to make room for the air conditioner when it was installed, can be pushed up even further by the burglar. Thus, he can crawl in over the top of the air conditioner without disturbing its mountings.

A dollar's worth of hardware can secure any window air conditioner. The raised lower part of the window should be firmly fastened to the upper member with angle irons on both sides (or more crudely, by pounding heavy nails most

Ordinary barrel bolt can be jimmied if the bolt does not extend far enough into the receiver, but it cannot be carded. Good security when knob is turned down into the barrel slot to prevent retraction.

of the way into the frame to stop the window from being raised any further, or even by nailing the window into fixed position). The top half of the window should also be nailed shut so that it can't be lowered, inasmuch as any catch originally put in as builders' hardware is inoperative with the lower part of the window raised.

Legs supporting that part of the air conditioner extending outside should be bolted to the sheet metal housing with the nuts inside the housing. The bolts themselves should either be recessed slotless carriage bolts or square-headed so that they cannot be turned when installed in a slotted leg. An air conditioner thus protected will *not* require a metal strap welded around the back of the unit and bent to lead inside where it can be screwed to the inner window frame, as some security-conscious folks have done.

Window-width ventilating fans are most often installed in an even more slipshod fashion than air conditioners. This is because many people take them out in stormy weather so they can close the windows against wind and rain. They should also take them out when the family leaves the house so they can close the windows against burglars.

Second Story Windows

Windows on the second floor, in particular, are often

left unprotected against intrusion. The term "second story man" is a well-established tag in police terminology. The ones who are particularly adept at acrobatics are known as "porch climbers." Many burglars specialize in going into houses through the windows on the upper floor because it is so easy, particularly when the homeowner conveniently leaves ladders lying around in the yard.

Windows on the second story, where window air conditioners and ventilating fans are even more widely used in bedroom windows than in the ground floor rooms, should be protected in exactly the same way as windows on the ground floor.

Sliding Glass Doors

Sliding glass doors, usually opening onto a patio or balcony, can be particularly tempting to burglars because they are often away from the street side of the house where a burglar can work unobserved by passersby. In the average sliding glass door, the entire sheet of glass can be lifted up into the top channel far enough so that the bottom edge clears the lower channel, allowing it to be lifted out. This should be feasible from the inside, to facilitate the replacement of cracked or broken glass. But thousands of glass doors are improperly installed "inside out," by people who just don't know any better; the glass can be lifted up and out *from the outside*. If the homeowner finds that his sliding glass doors have indeed been put in that way, he can put screws into the top of the door to block the channel, although taking the door out and turning it around is better. A properly installed sliding glass door always slides on the interior side of the stationary glass.

When sliding glass doors are shut, the space between the edges where they meet is usually big enough for a burglar to insert a wire to undo the simple catch with which most sliding glass doors are equipped. The homeowner can prevent a sliding door from being opened, even if it has no catch at all, by dropping a length of broom handle into the bottom channel, cut to fit snugly. A screw head (knob) next to the wall end of the broom handle will enable the householder to pull out the stick when he wants to slide the door open. People fussy about their decor can have a steel bar cut to the proper length to block the slide-in channel, instead of a broomstick. The same idea can also be used on the sliding windows, like trailer windows, now being built into many newer homes, as in the "high off the floor" bedroom windows that provide for more versatile furniture arrangements. These are dangerous windows because small children and elderly people can't get out of them in case of fire, but they can be entered with ease by a wiry burglar.

The Charley Bar made by Schlage for sliding doors is no more effective than the broomstick, but it is mounted chest-high as a brace between the sliding glass door and the wall, where a potential intruder immediately can see that the door is barred before he attempts entry. Some of the cam locks made by Corbin Lock Company, Berlin, Connecticut, can also be adapted to lock sliding glass doors, in both disc (wafer) cylinder and pin cylinder styles, provided that there's enough metal on the door frame to attach the hardware solidly. Most good locks for sliding doors are designed for commercial rather than for residential use.

Garage Doors

What a burglar can steal from the average family's garage, starting with the car, is often worth more than what he could steal from the house itself. Thus, garage doors should have locks as good as the doors to the house. The overhead doors used on most garages, sliding up into tracks, are usually equipped with center-keyed twist-grip or T-handled locksets—notably susceptible to a burglar using a pipe wrench.

A safer system, if not as convenient, is a keyhole that is

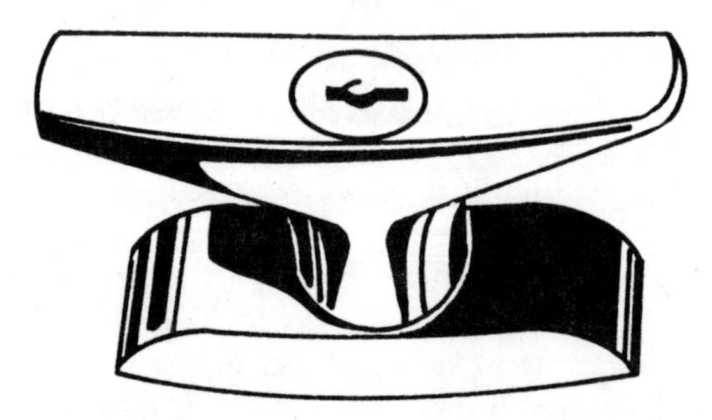

T-handled twist-grip lock should be considered a privacy lock, not a security lock; it is too easy for an intruder to open simply by twisting the whole thing off with a pipe wrench.

completely separate from a couple of "drawer pull" handles. Even then, an interior pin or bar system to block the tracks is a good precaution. And nobody, of course, should ever leave the garage door open when the car is being used (even "just to run over to the grocery store"); doing so is an open advertisement that there's nobody home to bother a pilferer.

Storage Enclosures

People who live in larger apartment buildings often have individual storage cubicles in the basement, provided by the management for tenants to store possessions that are infrequently used. A petty thief can raid a whole tier of such storage compartments in a matter of minutes, making off with suitcases, seasonal sports equipment, and anything else he can get any money for or can swap for narcotics.

Bins made of chicken wire cannot even be considered in terms of security, and locking them up only serves to keep honest people out. If the thief didn't happen to bring a pair of electrician's pliers to cut the wire, he only need grab a handful of wire near the edges and yank it loose from its staples.

Bins made of wood are usually about as sturdy as an orange crate, and prying boards off can be done by a child with a claw hammer. The least the supers should do when they build such storage areas is to nail the boards to the 2"x4" framework from the inside, so that the boards cannot be pried inward, but most take the easy way and tack the slats on from the outside. When the tenant buys a padlock for a contraption like that, he might as well be content with the cheapest one he can get, on the theory that there's no point in a knight wearing a helmet when he is otherwise naked.

Padlocks

Most people use a good many padlocks around the house—on basement storage compartments, lockers, tool cabinets, cellar doors, gates, and for bicycles and motorcycles. Inasmuch as the business is very competitive, price is just about as good a way as any to judge the dependability of any padlocks being considered. Good locking mechanisms cost money to make, and no padlock costing less than three dollars or so can be considered very good protection.

Cost-conscious lock buyers, who may have members of the family who are careless with their keys, might also keep in mind that single-bitted keys cost only about half as much to duplicate as double-bitted keys that have to be cut on both edges of the key blank. Any key blanks made of steel, as used for many foreign-made locks, also cost more when duplicate keys are needed for the locks.

The most famous are Yale padlocks, which have been on the market for over 100 years. In 1844, Linus Yale, a leading manufacturer of bank locks, patented a lock somewhat outside his chief line of business. His son Linus, Jr., although never associated with his father, having built a successful career as a portrait painter, nevertheless got interested in the 1844 patent, which he improved and perfected in his patents of 1861 and 1865.

This is still the basic lock, which, with some later modifications, means "residential security" all over the world as the Yale Cylinder Lock. The name "Yale" on a padlock is so impressive that a number of counterfeits have turned up, notably in foreign countries. Some with weird mechanisms never seen in this country are nevertheless prominently stamped "Yale."

Modern Yale padlocks (the real ones) include laminated warded locks, disc cylinder solid-body padlocks, and dial combination padlocks, all in various sizes and styles. A five-pin Yale cylinder lock with a 2-inch case and a

Laminated padlocks, with the body made up of layers of plates riveted together, appear to be as formidable on cheap models as on good ones; what makes the difference in the quality of locks is the interior mechanisms, not the construction of the bodies.

3/8-inch shackle costs $6.95. Quality has degenerated steadily over the years, and the locks now made by Eaton, Yale & Towne are no longer considered the best by most locksmiths.

Today the world's largest manufacturer of padlocks is the Master Lock Company, Milwaukee, Wisconsin. The company specializes in laminated bodies, made up of layers of plates riveted together, used for warded mechanisms as well as for four- and five-pin cylinder locks. They use a four-pin cylinder even in one of their "shell" padlocks, although most of their shells do indeed have only warded mechanisms. One of them, incidentally, the BRK model, is dangerous for a householder to use for security purposes if he gets hold of one accidentally; it has a specially constructed shackle purposely built so that it will shatter easily when given a good rap with a hammer or wrench (these locks are specified by fire insurance companies for use on sprinkler shut-off valves and fire hose housings).

The body of a wrought steel (shell) padlock is comprised of two pieces of metal made in a punch press and riveted together. It is quite inexpensive, but the two sides can be pulled apart by a strong man using two pairs of pliers or can be caved in with a hammer. Shell padlocks primarily are used as a means to restrict access, such as keeping children out of a tool room.

A solid body made of a single piece of steel or brass costs more to hollow out for the mechanism, but it's a lot stronger even though it can be cracked with a heavy hammer blow. Mechanisms in solid-body padlocks cover the entire range of locksmanship—warded, disc (wafer) cylinder, and pin tumbler—with respective costs depending on size as much as anything else.

Laminated padlock bodies can withstand pounding on the sides that would smash other locks because, in effect, each plate must be attacked separately. However, inspect the lock carefully before buying; a laminated padlock with

Shell padlocks, made of metal halves in a punch press, offer little reistance to anybody armed with a hammer or even a pair of pliers.

a hard-to-pick pin cylinder mechanism looks exactly the same as a laminated lock with a cheap warded mechanism.

Warded padlocks, the prices of which start as low as twenty-nine cents, can be picked with a bent hairpin because the tip of the key itself simply pushes apart the spring that holds down the shackle—and a thin metal pick simply avoids the wards as it goes in. The major uses for such padlocks are on locked-up property of limited value—oil tank caps, well covers, duffel bags, sheds for yard equipment, paint lockers, and so on. Because of the relatively large clearances between operating parts, warded locks are often preferred for applications where

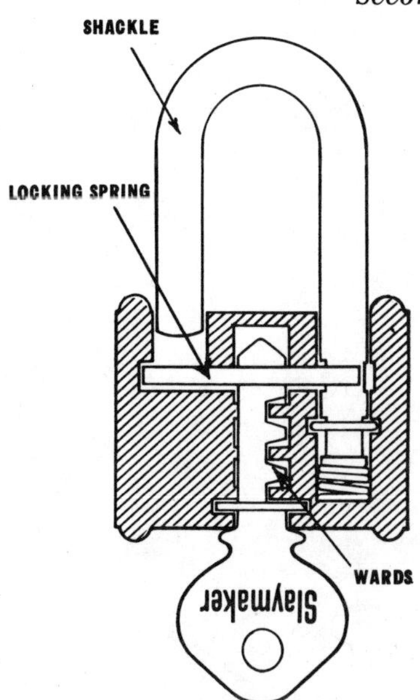

Warded padlocks can often be opened with a bent hairpin; all that's needed is to push away the locking spring.The wards, which can prevent an unauthorized key from turning in the keyhole, can simply be avoided by going straight to the locking spring.

sand, water, ice, or other contaminants could get into a lock.

Combination locks are also easy to "pick." Anybody with a little practice can hold enough pressure on the shackle to feel the drop as each of the three inner discs turns into proper place. (The only advantage to a combination lock is that the owner doesn't need to bother with carrying a key. However, many people who are prone to forgetting the number write it down and carry the notation for use when needed; they might as well be carrying a key.)

Padlocks with tumbler cylinders, though, are harder to pick than cylinder locks used in doors. This is because the intruder needs two hands to pick a cylinder lock: one hand

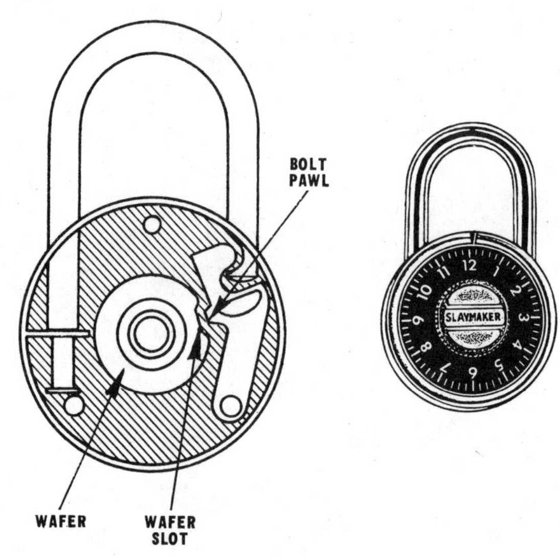

Combination locks are great for schoolchildren who keep losing their keys, but they can be opened quite simply by anybody with a good ear listening for the wafer slots to fall into place.

to hold pressure on the cylinder wrench and the other hand to manipulate the pick as he works on the tumblers one at a time. This is not easy to do with a loose-hanging padlock.

A disc cylinder lock is somewhat easier to pick because the wafer tumblers retract into the cylinder to enable it to turn, with a more positive feel of release. The tumblers in a pin cylinder lock must be pushed up *out* of the cylinder to release it for rotation, and in a five-pin lock the pick must be used several times on each two-piece pin to get the separations all lined up evenly along the top of the cylinder (see Chapter 1). This process takes a steady hand and is not well adapted for the manipulation of padlocks.

The householder should beware of coded padlocks. These code numbers are for the convenience of locksmiths in case the owner loses his keys; the number tells the locksmith what kind of key is needed, and how to make the

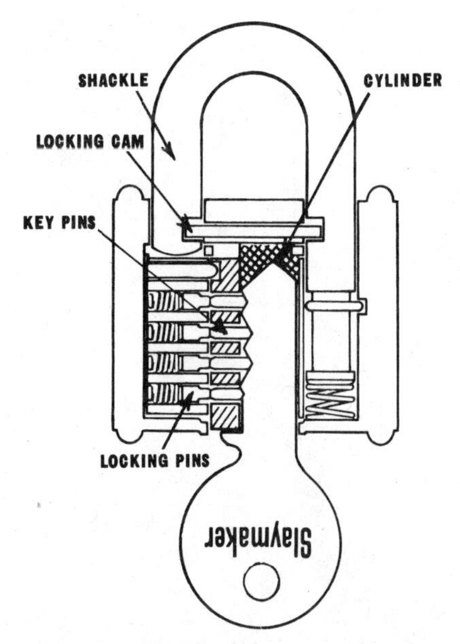

Pin cylinder padlocks are difficult to pick because two hands are needed for the operation with a separate burglar tool in each hand.

duplicate. But burglars can get hold of these codes, too (we did, merely in researching this book). Any householder with such a padlock should make a record and then file off the numbers or symbols. Many makes now have the number printed on the padlock with removable ink, or on a detachable tag.

There are almost always easier ways to attack closures protected with padlocks than with cleverness. A good crack with a hammer, for instance, will open many padlocks, including some of the expensive ones costing as much as eight dollars each. By placing the top of the shackle against the solid surface of the hasp, a sharp blow on the bottom of the lock will smash the internal mechanisms in many cases.

Some of the most shock-resistant padlocks on the market are made by the Junkunc Brothers American Lock

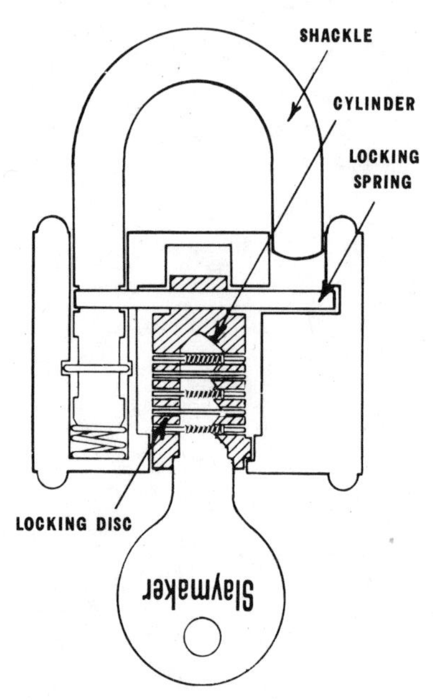

Disc cylinder padlocks with wafers protruding into the shell are easier to pick than pin cylinder locks because the discs have a more positive feel of release.

Company, Crete, Illinois. Instead of using internal levers that can shear under pressure, the American padlocks have a solid steel ball wedged into each side of the shackle within the lock body. The more pressure that's applied, the tighter they wedge. The six-dollar American disc cylinder lock with a 7/16-inch shackle is as secure a lock as most people ever need, but the company also makes a padlock with a shrouded shackle that is virtually indestructible.

Along with regularly keyed padlocks, Junkunc-American also makes seven-pin tumbler padlocks with the pins arranged in a circle to foil would-be lock pickers of cylinder locks, using a tubular key like the Chicago Ace cylinder lock (see Chapter 1). The configuration allows for over 800,000 key changes. The Dynation Corporation,

Little Falls, New York, also makes a comparable padlock using a tubular key. Unlike the Chicago Ace tubular keys, these tubular padlock keys can be duplicated by locksmiths, and are of a slightly different size.

Slaymaker Lock Company, Lancaster, Pennsylvania, makes a maximum security lock using a locking ball on each side of a shackle made of heat-treated molybdenum. S. R. Slaymaker started the company in 1888 with the idea that padlocks could be made on a mass production basis, instead of one at a time by skilled locksmiths, as theretofore. Today the company makes the most complete line of padlocks in the world, with most of its distribution through volume outlets such as dime stores, discount hardware chains, and mail order houses.

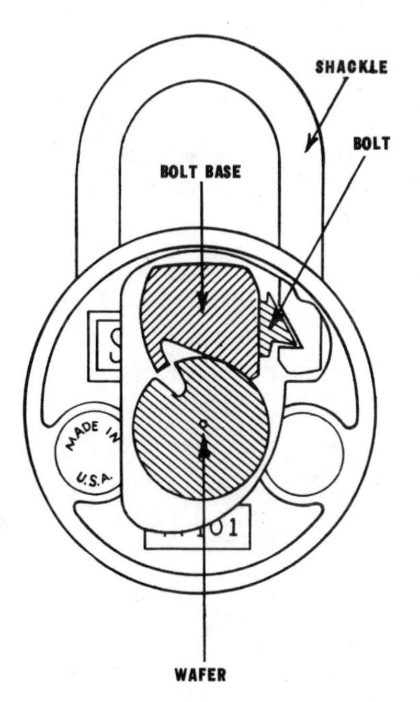

Padlocks with "arrow head" bolts that lock the shackle can be snapped open with a sharp blow.

A shrouded shackle makes the American-Junkunc hardened steel padlock virtually indestructible, and even a sledge hammer won't knock it open.

Montgomery Ward, incidentally, sells one padlock (among others) that's better than it needs to be. The $10 Ideal has a sixteen-pin disc mechanism in a solid brass body, and it is virtually pick-proof. Like all good locks, including many in the $3 to $4 price range, Ward's Ideal locks on both sides of the shackle.

Some padlocks cost as much as $30, including top-of-the-line models made by Schlage, Sargent, and Best. Their stainless steel and brass pin cylinders are extremely durable. But some burglars carry heavy-duty cable cutters (up to four feet long) that can cut any padlock shackle with one snap. To counter this hazard, the American Lock Company makes a hasp lock on which the shackle can't be cut because

A padlock with a "burglar alarm type" mechanism, using a tubular key, is about as pick-proof as anybody needs in a padlock application.

it doesn't have one. The six-pin tumbler Superlock is a massive piece of hardened steel (3-1/16" diameter × 1-9/16" thick) that fits over and completely covers the area where the back of it locks into the hasp.

More frequently, the hasp is attacked instead of the lock itself, because the hasp's staple is often thinner than the lock shackle and can be cut with just about any bolt cutter. Sometimes the entire hasp, if installed with small screws, can be jimmied off, unopened expensive lock and all.

Frequently the physical circumstances are such that the mounting screws cannot be covered with the hinged hasp when it's closed, which would ordinarily allow anybody with a screwdriver to take off the whole works. In such cases, the householder should use non-retractable screws,

The American hasp lock's shackle cannot be cut because it doesn't have one. Six-pin Superlock fits over and completely covers the hasp as it locks on.

which must be drilled out for removal. They can be screwed in as easily as any other screw, but opposite halves of the slot in the head of the screw are machined at a slant so that a screwdriver can't get a grip in trying to remove such a screw counterclockwise.

The best way to mount security hardware is to use slotless-headed carriage bolts going all the way through the wood, with the nuts on the inside. But not even this will stop a determined professional with the right burglar tools if he has enough time.

Vacation Risks

A burglar can take all the time he needs when the family is away for vacation. He can scout the entire premises for the easiest locks to get through when he knows that nobody will be home for a couple of weeks. Some families are even foolish enough to brag through the newspapers when they are leaving for a trip to the Bahamas or taking some other trip they hope will turn all their neighbors green with envy. Thorough society editors even print the date of departure and the travelers' address if given such information.

Vacationers should shut off newspaper and milk deliveries, have a neighbor take in the mail, hire somebody to mow the lawn or shovel the snow, and use an electric timer that will turn lights and a radio off and on. Small automatic timers, which simply plug into any standard wall outlet and reset themselves every twenty-four hours, cost as little as $6 or $7. Even a light-sensitive photoelectric switch, to turn on yard lights automatically at dusk and off again at dawn, sell for only $10 or $15.

Simply leaving a bathroom light turned on all the time is an *announcement* that nobody is home, if it is noticed for several nights in a row, particularly if noticed burning during the day. The best timers turn lights on and off at irregular times, so that no pattern can be discernible even to somebody casing the place.

One of the best appliances to hook up to the timer, along with lights, is a television set. Even a radio playing in a lighted house will give most burglars pause, but a television set in operation is a "sure" sign that somebody's home because "nobody" ever leaves the house with the television on.

Some people even leave the phone off the hook whenever they are away. Thus, when a would-be burglar attempts to check up on their really being away by making a fake phone call, he gets a busy signal instead of a series of unanswered rings that would be a tip-off.

A house that looks lived in, with all means of entry securely locked, is safe from over 90 percent of all burglars.

3

Surveillance

"This damn woman, whose door I had pushed open, didn't start yelling 'Help!' which just gets most people to slam their doors so as not to get involved. Instead, the bitch started to scream 'Fire!' and she practically had the whole neighborhood there in a matter of seconds"—Convict at California State Prison, San Quentin.

Robbery and rape, as well as burglary, are also on the increase. Robbers who force their way into a house are often people who, for a variety of reasons, are oblivious to the chances they are taking, compared to the safer crime of burglary in which identification is seldom possible. However, most burglars, if confronted accidentally, then become robbers even if they hadn't intended to be.

Nobody should ever let anybody into the house if the visitor is not known. This includes any strangers who ask to use the phone because their car has broken down, or who say that they have been robbed and want to call the police.

Make the phone call *for* them, but don't let them in. This is admittedly not being much of a Good Samaritan, but it's a lot safer, especially for anybody who is in the house alone. As for meter readers, city inspectors, poll takers, people who say that they are running for political office, door-to-door salesmen, or anybody else who may or may not be what they appear to be, ask them to show identification before inviting them inside.

Door Chains

Any householder who does not have a window located in or near the door so that he can look out to see who is at the door would be well advised to get a chain guard for the door. This device is intended to allow the door to be opened a few inches but no further; the chain can be unhooked only when the door is closed.

That's the theory, but in practice many door chains are so long that somebody outside *can* reach in and unhook the door chain from its slide. Others are so skimpy that they can be cut with an ordinary pair of pliers. Most of the door chains sold in dime stoes are sold complete with screws—some of them as small as half an inch long. A good shove will rip them right out of the wood where the chain is fastened to the frame. The forty-nine-cent ones are made of metal so light, stamped out of a punch press, that they can be bent even accidentally. Such gadgets are called "wrought steel," but they are too tinny to be considered dependable security hardware.

Even some of the better door chains have only two screws fastening the chain holder to the jamb next to the door, instead of the four they ought to have. A good door chain holder should have the screws going all the way through the jamb and on into solid wood. Both the slide and the chain holder should be solid formed or extruded brass, and the chain should be case-hardened steel with welded links. Door chains of this quality are usually in the three- to four-dollar class.

Some models are made to unlock from the slide only with a key—dangerous to have in case of fire when people have to get out fast. Others can be locked and unlocked from the outside. Wessel Hardware Corporation, Philadelphia, Pennsylvania, makes a ten-dollar model that sets off a loud built-in alarm when pressure is exerted against the chain. The purpose of all of them is to see who is at the door while keeping the door mostly closed, and none of

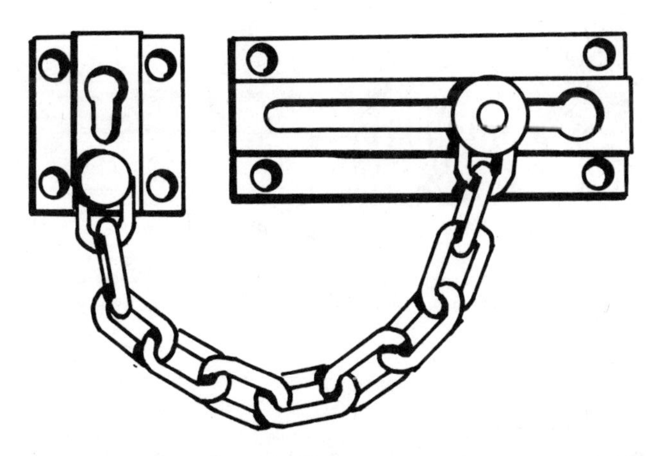

Chain guard should have four screws on each side, and should be made of extruded brass for maximum security.

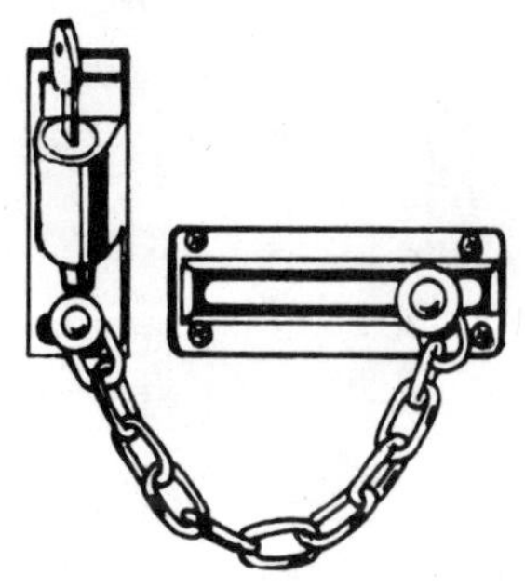

Chain guards that open only with a key serve no practical purpose—and are dangerous in case of fires.

them are a satisfactory substitute for a good deadbolt when meant for security.

Viewing Lenses

A fish-eye viewer is a better and safer means of surveillance than the chain guard for the owner of a solid door, effectively allowing the householder to see out without being seen. A wide-angle lens in the one-way

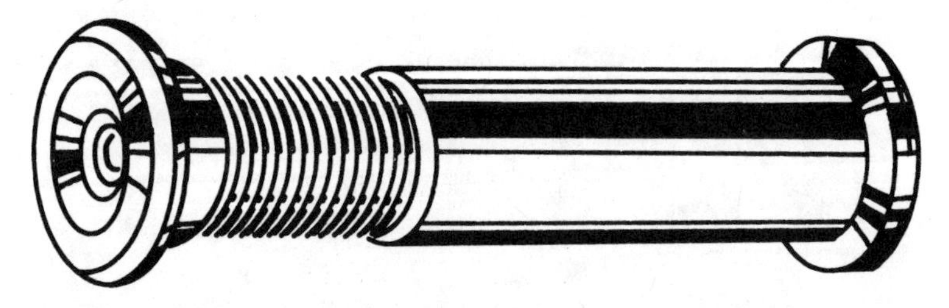

A viewer mounted through a solid door, presenting a wide-angle "fish eye" view of the hallway outside the door, is cheap, effective, and easy to install.

viewer provides a full 175-degree circle of vision.

Mounting it in the center of a door at eye level is a simple matter of drilling a half-inch hole through the door, sticking the viewer in like a plug, and fastening it on the inside with a screwdriver or edge of a coin. In most areas, a fish-eye viewer for average doors (1-3/8 to 2 inches thick) costs less than three dollars.

Baldwin makes a whole line of door knockers equipped with fish-eye viewers, which the company calls Observ-O-Scopes, in some twenty different styles and sizes. When the viewer is inconspicuously set into a door knocker, the householder peeking out is less likely to see just an eyeball looking in at him.

Intercoms

For apartment buildings, the safest of all surveillance systems is to have a doorman, or even a security guard at a reception desk, to announce visitors. But that is strictly for luxury buildings, and the average apartment dweller considers himself lucky if he has a responsible super around the premises. All tenants, though, should insist on having an intercom system in working order, whereby they can talk to anybody ringing their bell from the vestibule before they press their buzzer button to release the door latch and admit the visitor to the hallway.

The intercom is plugged up or otherwise inoperative in thousands of apartment buildings, and all an intruder need do is ring apartments on the top floor to gain access and then rattle the doorknobs on the lower floors until he finds one that's open. Such buildings might as well not have a "security" door at the building entrance at all.

It is downright amazing to see how many people press the vestibule latch button without checking to see who is ringing the doorbell, even when the intercoms do work. They just don't bother. In a building with, say, twenty-four apartments, anybody who rings all twenty-four doorbells next to the mailboxes will almost always get three or four tenants to open the hallway latch.

The vestibule's door to the hallway is necessarily equipped with a cylinder lock that can be opened by all the tenants' housekeys, a lock often constructed so that it can be opened by anybody using just the edge of a dime as a key. The security for tenants in such buildings is practically nonexistent except for the locks they personally have put on their individual apartment doors.

Closed Circuit Television Systems

Closed circuit television (CCTV) is used in a growing number of apartment buildings, particularly in quality dwellings. When this electronic surveillance equipment has a camera with a wide-angle lens, it can cover the mailboxes and doorbells as well as the entrance door. A standard CCTV camera lens covers an area of about five feet at a ten-foot distance, and a wide-angle lens covers an area twice that big.

CCTV is not as complicated as is commonly supposed. The camera has no transmitter as such but is simply hooked up to the television receiving set (monitor) with a coaxial cable. A regular television set, no matter how sensitive or how expensive it is, cannot tune in on the picture sent by

Closed circuit television surveillance equipment ranges from expensive, centrally monitored systems to fake cameras that only get potential intruders to *think* they are being watched.

a CCTV camera unless the set is hooked up directly to the camera's cable.

The cost of CCTV systems largely depends on the amount and type of equipment used. In one system, a single camera is connected to a monitor in the superintendent's quarters, where he can keep an eye on vestibule activity on a selective basis with spot checks made at times known only to himself. The Panasonic (Matsushita Electric Corporation of America) mini-system with a nine-inch video screen on the receiving set, is well suited to this type of system and costs under $500 complete. If the building has more than one entrance, up to three cameras can be added at a cost of

$180 each. These additional cameras are monitored on the single receiver by turning a selector dial that works on the same principle as the channel dial on an ordinary television set.

In some CCTV systems, each tenant has monitor equipment of his own; when operated in conjunction with an intercom system, the tenant can see as well as hear his visitor. If the building owner is willing to use more expensive cameras that include audio as well as video capabilities, the camera can be cabled to each individual apartment in such a way as to enable each tenant to use regular television sets instead of separate monitors. Just by switching to the "cable channel," the tenant can get the picture from the vestibule. The total cost of such systems can be surprisingly low—perhaps a few hundred dollars per apartment.

Anybody with CCTV equipment is much more likely to make full use of the surveillance equipment than people who have only voice communication with the vestibule. Tenants who are dependent on an intercom really have no way of telling if the person downstairs is really who he says he is; but when they can see him as well as hear him, they are less likely to push the latch-release button automatically any time the doorbell rings.

An intruder is also more chary of blithely pushing all the doorbells with the expectation that somebody will be careless enough to let him in without checking, especially if he knows that any number of people may be watching him. The very presence of a CCTV camera trained on him is enough to prompt the average no-gooder to go elsewhere for whatever nefarious business he has in mind.

The deterrent value of CCTV is so well recognized that a growing number of firms entering the security industry now sell *dummy* CCTV cameras, complete with fake lens and line cord. Most have an electric light that glows red to signify that the camera is "on," and some dummies are even built as scanners, with a small electric motor moving the camera back and forth in a predetermined arc just like the real ones.

Dummy CCTV cameras, with unconnected exterior knobs, switches, lens controls, and all, sell for as little as $45. Ademco (Alarm Device Manufacturing Company, Syosset, Long Island, New York) sells an exact duplicate of its truly efficient Model 1501 camera in a dummy version, the Model 1503, with everything the same on the outside but with no working parts. Security Systems International, Salisbury, Maryland, sells dummy duplicates for about one-eighth the price of its expensive CCTV cameras.

To be really effective, any dummy CCTV camera must look like a *standard* unit; it should not be distinctive in its design in any way. One dummy manufacturer had the poor sense to prominently put his nameplate on the fake camera; as soon as word got around that he made only dummy cameras, with no line of real equipment, thieves went around and stole his installations left and right from the misguided buyers who thought that they were buying protection.

Of course, no system of surveillance is totally foolproof. A can of spray paint will "blind" a CCTV camera lens while a job is being pulled, but that's a ploy used only in such big hauls as bank robberies when the crooks are willing to take long chances.

In all cases of residential security, thorough surveillance is one of the most effective of all crime deterrents. Some people decry many types of surveillance as an invasion of privacy, and in some cases it *can* be when it is misused. The very word *surveillance* has a connotation of not trusting whoever it is you're keeping an eye on. But that's one of the prices that has to be paid for today's social climate. If surveillance is what it takes to keep malefactors from pulling any rough stuff, that's the way it has to be.

4

Firearms

"When I find out that somebody keeps a gun in the house, I make a special effort to get in and steal it because there's such a good market for guns in Chicago"—Prisoner, Cook County Jail, Chicago, Illinois.

This book is not a proper forum for entering the debate on gun legislation and gun control. But the simple fact is well established that many homeowners do buy firearms as a security measure. These are the people for whom this chapter has been written.

Certainly, all handguns should be registered—and kept out of the hands of junkies, drunks, psychotics, convicted felons, persons with a record of emotional instability, and children. But even for the responsible person who wants to keep a gun to protect a home and family, there are certain human factors and legal risks that should be borne in mind.

There are at least three minimum requirements for owning a gun: (1) the owner must know how to use it; (2) all family members must be taught to treat it with the respect it deserves; and (3) the owner must be prepared to deal with the fact that he may have to kill somebody if the circumstances make drastic action with a deadly weapon necessary. This third requirement generates the most heated debate, but it is, perhaps, the most obvious.

Anybody who puts a gun on a burglar must face the possibility that pointing a gun at him may serve to provoke him into an attack of his own, and the homeowner must be prepared to retaliate against any such violence. People who would hesitate to shoot would be better off just saying, "Tell me what you want and I'll give it to you . . . it's all insured anyhow."

Nobody should *ever* point a gun at anybody unless he has the full intention of killing him at the first threat of resistance. To do otherwise would be dangerous; if you only wound an intruder, you have given him further provocation to kill *you*.

If the householder gets the drop on a burglar who does indeed freeze on command, he should make no move to frisk him or otherwise get any closer than eight feet from him until the police arrive and take over. Until the police arrive, the burglar should be ordered to stand with his face against a wall with his arms outstretched. If he refuses to do so, the person with the gun should be prepared to fire.

Legal Risks

Never shoot at anybody unless he is inside the house. Shooting a burglar either before he gets in or after he gets out can result in all kinds of unpleasant legal consequences. If he is shot before he gets inside, his lawyers can claim that he hadn't meant any harm—and the burden of proof is on the prosecution. Just prowling isn't illegal beyond simple trespassing—and there's nothing else to charge a burglar with unless he has indeed broken in. If he is shot while running away after having broken in, the householder could have a tough time disproving the burglar's claim that he had never been inside.

Judges in many states take a dim view of anyone shooting anybody just to protect property. If the burglar survives, he is justified in the strict eyes of the law in suing the householder who shot him in jurisdictions where the use of deadly force is considered justifiable only when "all other means of preventing the crime have been exhausted." In many eastern states, the only reason anybody can legally shoot another person is when his own life has been threatened. In all such cases, the homeowner should get a lawyer and let the lawyer do the talking.

Finally, never buy a handgun "on the street." Get it from a reputable dealer. The chances are excellent that street pistols are stolen or even have been used in a shooting. Any used gun, for that matter, may be in such poor condition that it may fail to work when it is needed.

Handguns in general should be considered as offensive weapons, never as defensive weapons in the strict sense. They are ill-suited for anybody trying to defend himself against anyone armed with a rifle or even a shotgun because their range is so limited and their barrels are not long enough for completely accurate sighting. Aside from target pistols, the only purpose of a handgun is for use in shooting somebody.

Automatic Pistols

The drawbacks inherent in automatic pistols make them less than ideal for use as a house gun. Their chief advantages are twofold: (1) they hold more rounds of ammunition than revolvers; and (2) second and subsequent shots can be fired more rapidly than from a revolver, unless the latter is fired double-action (in which case it is quite difficult to hit anything more than a few feet away). Cocking a revolver, so that it can be fired as a single-action pistol, takes more time than just continuing to pull the trigger repeatedly with an automatic, in which the recoil from the first shot throws back the slide, ejects the spent cartridge case, and feeds the next cartridge from the clip in the handle into the firing chamber as the action is recocked automatically.

But this advantage of an automatic is also its chief disadvantage—it is *too* automatic. Leaving a live cartridge in the chamber, which means that the gun is fully cocked, is just a bit too dangerous for a house gun. Only the best automatics, such as the top-line Colts, have a manually operated hammer that has to be thumb-cocked for the first shot. Otherwise, an ordinary automatic is fully as dangerous as leaving a hammerless double-barreled shotgun around the house fully loaded, which also cocks the shotgun's action with nothing but the safety catch to prevent accidents. Most loaded and cocked automatics can go off if accidentally dropped on a hard surface.

Keeping an automatic pistol fully cocked for long periods of time can also weaken the springs, which, if kept constantly under tension, might not be strong enough when needed to actually fire the cartridge. The alternative is to keep all the cartridges in the clip, with the chamber empty (although keeping the magazine fully loaded at all times can eventually weaken its feeding spring enough so that it won't snap the cartridges up cleanly, and the eighth or ninth cartridges may jam the whole action).

The empty firing chamber, however, means that the owner has to activate the pistol by pulling back the slide to get a cartridge into the chamber, a two-handed operation that takes a certain amount of muscle and time, and also makes an alarmingly loud sound in a quiet house. If for any reason the user doesn't have the use of both hands free, he is out of business if the cartridge doesn't fire because it has corroded or if—as happens—it is faulty because of a factory defect. If a cartridge in an automatic doesn't fire, the user has no second chance to just pull the trigger again.

The main reason criminals prefer automatics over revolvers is because the flat-sided automatic doesn't make as big a bulge when carried as a concealed weapon. That moot advantage is not enough to offset still another limitation of automatics—the ammunition. Caliber for caliber, bullets fired from an automatic have less stopping power than the soft-nosed bullets that can be used in a revolver. Automatic cartridges have great penetrating power, but that only means that they will ricochet all over the place. Soft-nosed bullets, which expand when they hit, cannot be used in an automatic because the spring mechanisms would deform them before firing.

The hard-coated bullets used in automatics are like the jacketed bullets used in military rifle ammunition, where accuracy at long range is a good feature, but of no use to anybody shooting across the width of his living room.

Although very few automatics are used by police departments, a number of homeowners still prefer them for the advantages they do have. The most dependable and most rugged automatic pistols for house use are made by Colt, a company that does not fool around with small calibers but makes such automatics only with heavier stopping power—9 mm, .38, and .45 calibers. Retail price is $175 ($17 more for a nickel finish, available on the heavy 39-ounce .45 Government Model). The sporting and target models made for .22 caliber cartridges, such as the Colt Woodsman, are not household guns; the bullet is too small to use on anything as dangerous as a man and the barrels are too long for easy storage and handling.

Smith & Wesson's Model 39 holds eight rounds of 9mm ammunition in the magazine, measures 7½ inches overall, has the same external cocking hammer featured on the Colt, and costs $219 or $143.50, depending on whether it has a blue steel or nickel finish. Basically the same but with a magazine holding fourteen rounds, the Model 59 costs $155 (or $169 with a nickel finish). Smith & Wesson also makes a .38 automatic over 8½ inches in length, the Model 52, at $233, but this target gun uses wad cutters only, in a five-round clip. The Smith & Wesson .22 target

Government model Colt automatic, firing a .45 bullet, is better for military combat than for household use.

automatic looks more like a zip gun than a real pistol, and it is in the same class as the Colt Woodsman.

Browning's standard .38 caliber automatic is $30 less than Colt's, but it is also sold loaded with engraving for $340 as the Renaissance Model. The standard 9 mm Browning is $175—$460 when fully engraved as a Renaissance version with adjustable sights (which no home owner needs because any shooting he does will be at extremely close range).

Harrington & Richardson's .38 Model HK-4 is only $90, but the compact automatic at 24 ounces is so light in weight that the recoil from a cartridge this size is enough to bounce a shooter's firing hand half a foot into the air every time he pulls the trigger. It also lacks a cocking hammer, and it does not have the safety grip a good automatic should have. This safety grip, a feature on both Colts and

Smith & Wesson Model 39 holds eight rounds of 9mm ammunition in the magazine and has good safety features.

Brownings, makes the trigger inoperative unless the handgrip is firmly grasped in firing position.

There are a number of pocket-sized automatics on the market, .25 caliber foreign makes, and even .32 caliber pistols that would fit in the palm of your hand. None of them have enough stopping power for practical use as a house gun. Most of the handguns made in Belgium and Germany use good quality metal alloys, but some of the guns coming out of Spain, in particular, are made of metal only a little better than cast iron. Now that Taiwan is also in the gun business, some of the handguns around are as dangerous to the shooter as to the object or person he is shooting at.

Revolvers

The most powerful handguns, which would be the

.38 version of the basic Smith & Wesson automatic is designed as a target pistol, shooting wad-cutters only from a five-round clip.

.357 or .44 magnum revolvers, do not make the best house guns. These brutal cartridges, which have an extra measure of powder behind the bullet, can penetrate the engine block of an automobile. They not only could go entirely through the burglar you shoot, but also could keep going right on through the wall behind him to wound or kill somebody not even involved. A .38 slug—.45 tops—is enough to discourage anybody if it hits him anywhere in the body, and it will most often stay in him.

Revolvers have simple mechanisms compared to the complexities of automatics and for this reason, seldom get out of order. Revolvers also cost less and can be fired double-action for speed or single-action for accuracy. They are a lot safer when kept in a nightstand drawer with an empty chamber under the hammer in the revolving cylinder; thus, the pistol cannot possibly fire accidentally if dropped onto the floor.

At one time Smith & Wesson made the best revolvers in the world, but that company's quality control has not kept up with that of other manufacturers in the industry. The main reason so many police departments use the .38 Smith & Wesson Police Positive is because it costs less than the comparable Colt.

One of the best guns Smith & Wesson produces is the .38 Centennial—which is designed for people who don't know much about handling guns. Most revolvers do not have safety mechanisms because there is no real need for a safety on a firearm unless the gun is carried at full cock. But the Centennial has safety features that make it almost impossible to fire accidentally.

One of its features is the same kind of safety grip found on a good automatic, which makes the trigger operative only when the hand grip is firmly grasped in firing position. As another safety feature, it is "hammerless," with no external hammer that could be pulled back by somebody just fooling around with the gun. Like all handguns with hammerless actions, the Centennial can be fired only double-action, with no possibility of cocking the hammer and then firing single-action for more accuracy.

The five-shot Centennial (Model 40) has a 2-inch barrel, weighs 19 ounces, and costs $103.50 or $113, depending on finish. The same five-shot in an Airweight version (Model 42) is far too light at 13 ounces for accurate repeat shooting, and costs $107 or $121. The comparable Model 38 Airweight Bodyguard "humpback" at 14½ ounces is $102 and $116, or $100 and $110 in the Model 49 20½-ounce version.

Law enforcement officers now consider Colt's line of six-shot revolvers, all made with external hammers, to be the most dependable for home defense. The .38 Colt Detective Special has just about anything a homeowner can

Smith & Wesson's five-shot Centennial Model 40 is for people who don't know much about handling guns. Grip safety prevents its being fired by children.

ask for in a handgun. The short 2-inch barrel makes it fast to use, it has a slanted ramp front sight that won't catch in clothing when drawn fast, and its 22-ounce weight makes it reasonably steady for a revolver. List price is $140 for the blue steel model, or $150 with a nickel finish.

The same thing, with an aluminum alloy frame that reduces the weight to 16½ ounces for easier carrying, is sold as the Cobra for $142 ($162 nickel). The grip on a Cobra is lowered behind the trigger guard to provide a more positive pull with the trigger finger, but a big man can get only two fingers around such a grip, with the little finger under the butt. With a full-height grip, the aluminum alloy .38 is sold as the Agent for $141, or for $146 with a shroud. Ease of carrying should be of interest only to a plainclothesman, and these lightweights are too hard to control under firing conditions for the average man.

Colt's Official Police Special is a .38 with a 4-inch

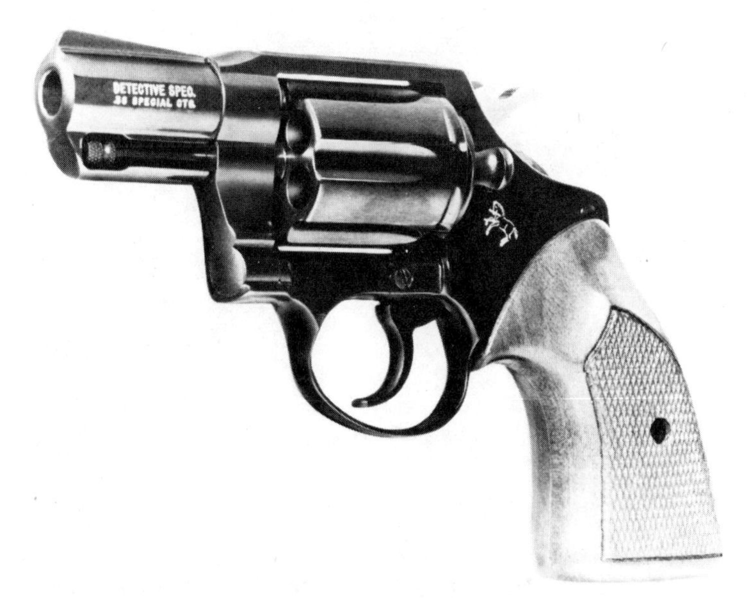

.38 Detective Special made by Colt has ramp sight for fast draw, short barrel for easy handling, and enough weight for accurate shooting.

barrel, it weighs 33 ounces, and it costs $147. The other Colt .38 is the Diamondback with either a 2½- or 4-inch barrel at $175 ($188 for the nickel 4-inch version). The 28½-ounce Diamondback has the same ventilated rib found on the more powerful .357 magnum Python and is most often equipped for target shooting. The 43-ounce Python is an ultimate in double-action revolvers and costs from $253 or $282 depending on the finish, in 2½-, 4- and 6-inch barrel lengths. It is more gun than anybody needs for shooting burglars. So are the .357 Lawman and Trooper models, with barrel lengths from 2 to 6 inches and costing from $150 to $200.

The only .45 caliber revolvers made by Colt are single-action replicas of historical handguns such as the Frontier Model, "The Gun That Won the West." These "cowboy guns" are collectors' items or are fun to take out to

Colt's only .45 revolvers are single-action replicas of historical handguns and are unsuitable as house guns.

a target range, but they are too slow and clumsy for a homeowner to use for protection.

The most powerful handguns are made by Smith & Wesson, which manufactures the biggest line of handguns —in all ways—in the industry (they make some 40 different models, most of them in different barrel lengths and finishes). The massive .44 magnum Model 29 is the kind of six-shot monster used by Clint Eastwood in the Dirty Harry movies, and is nothing for the average homeowner to fool around with. It costs $215 for either a 4- or 6½-inch barrel, or $221 with an 8-inch barrel, with choices of bright blue or nickel finish.

Smith & Wesson's .41 magnums are the standard six-shot military and police Model 58, at $118 or $127, and the Model 57 with precision sights and a target grip at $214 in 4- or 6-inch barrel lengths ($221 with 8½-inch barrel),

.44 magnum made by Smith & Wesson has more fire power than the average homeowner is ever likely to need.

all with choice of finish in nickel or bright blue.

All five of Smith & Wesson's .357 six-shot revolvers have adjustable rear sights, but the $151 round-butt Model 19 with a barrel only 2½ inches long is definitely not in the target class. The Model 19 is also available with a square butt and 4- or 6-inch barrels. The Model 27 has 3½-, 5- or 6-inch barrels in nickel or blue finish at $179, or at $185 with an 8½-inch barrel. The workmanlike Model 28 is available in blue finish only, with 4- or 6-inch barrels at $127, or $134 with target stocks. The Model 66 .357, at $174, is a showpiece with a satin-finish stainless steel frame. Smith & Wesson also makes a .45 target pistol, the Model 25, which has a blue steel frame, a 6½-inch barrel, and a $165 price tag.

The standard six-shot Smith & Wesson .38 revolvers undersell their .38 Colt counterparts by quite a bit. The popular Model 10, with 2-, 4-, 5-, or 6-inch barrels, costs

Smith & Wesson's .357 can shoot through the engine block of an automobile . . . enough fire power for any homeowner.

$96 or $105, depending on the finish, with rounded or square butts. The special "Heavy Barrel" version is available in a 4-inch barrel only, with a square butt. The Model 64 at $127 is a showpiece edition, with either a regular or a "Heavy" 4-inch barrel and the satin-finish stainless steel frame.

The Model 67, at $143, is the same as the Model 64 except that it is fitted with adjustable rear sights. So is the six-shot Model 14, a .38 with a 6-inch barrel at $118, or at $124 with an 8½-inch barrel; it is even made as a single-action target pistol at $136 and $142.50 for the different barrel lengths. Model 14s are available only in blue steel finish. The Model 15, precision sights and all, is made in 2- and 4-inch barrel lengths with a choice of finishes at $122 and $128 respectively.

The main reason Smith & Wesson .38 revolvers are used by so many police officers is because they cost less than the Colt counterparts.

In pocket revolvers, the Model 12 is an Airweight .38, weighing only 16 ounces with a 2-inch barrel (it is also available with a 4-inch barrel). It costs $102 or $116, depending on the finish. The Model 37 is also an Airweight (14 ounces), but drops down to a five-shot cylinder as the Chief's Special, with 2- or 3-inch barrels, tagged at $98 for the blue model or $107 for nickel. The same thing at regular weight—19 ounces with a 2-inch barrel—is the Model 36 at the same prices. The showpiece Chief's Special, with a 2-inch barrel on the satin-finish stainless steel frame, is $127.

Smith & Wesson's .32 revolvers are the hand ejector Model 30 and the "Regulation Police" Model 31, both of them six-shots with 2-, 3-, or 4-inch barrels and priced at $96 and $105 according to the finish. All Smith & Wesson revolvers have swing-out cylinders.

The Smith & Wesson line includes seven different six-shot .22 revolvers, all fitted with adjustable rear sights

Weighing only 16 ounces, the Smith & Wesson .38 pocket revolver is difficult to fire with any degree of accuracy.

for plinking, with barrels from 2 inches (!) to 8½ inches in length, with prices from $110 to $175. Except for the Models 51 and 53, which fire .22 magnum cartridges, nobody but a highly skilled marksman—and calm at the time—would dare fire one of these pistols at anybody attacking him.

There are a great many more makes of revolvers around than automatics because a revolver is so much easier to manufacture. They include all the Saturday Night Specials that figure so largely in the FBI's National Crime Index. Around the turn of the century Sears and Ward's, among others, were selling revolvers for as little as 60¢—and some of them are still around. Modern versions, selling for $15 or $20, have little in the way of improvements. Many are imported counterfeits or quasi-counterfeits hopefully stamped "Cult" or "Smythe & Weston."

Some of the most dangerous of the cheap handguns are

those with top-breaking actions; that is, the pistol is hinged in front of the trigger guard and opens at the top in front of the hammer for extracting spent cartridges and reloading. A revolver such as a Colt with a solid top frame, with a cylinder that swings out from the side for reloading, has much less chance of misfiring. Top-breaking revolvers can all too easily blow open the latch, and a shooter using powerful modern smokeless ammunition (thousands of pistols were made to withstand the forces of only black powder) can blow his hand off.

The best of a bum bunch are made by Harrington & Richardson. Their barely minimal .32 six-shot Model 632 has a cylinder that falls out in your hand completely when the holding pin in its front center is pulled out. This even has an advantage—the user can insert a fully preloaded spare cylinder, instead of taking the time to clean out and reload the original cylinder. Retail price is about $55.

Harrington & Richardson's other .32 six-shot revolvers are the 732 Model, with 2½- and 4-inch barrels, at $61.50, and the nickel-plated Model 733 with a 2½-inch barrel at $68.50. The 732 and 733 models have swing-out cylinders, just like on a good revolver. Along with a number of .22 plinkers, Harrington & Richardson also makes a five-shot .38 with a 2½-inch barrel and a break-open top for reloading. Retail price of this Model 925 is $75.

Shotguns

One reason there are only a third as many burglaries (per 1,000 inhabitants) in rural areas as in big cities is because most farmers own shotguns and know how to use them. A 12-gauge shotgun is an effective weapon even in the hands of the inexperienced because it needs to be pointed only in the general direction of its target when fired. Whereas a handgun or rifle has to be aimed accurately to hit anything with the single bullet fired, the charge from a shotgun covers a wide area.

The 15 ⅓-inch lead balls fired from a 1 7/8-ounce charge of 00-buckshot can all but tear a man apart if he is hit with the full load. Even a hit with only one or two of them can maim or kill. Each slug in a 00-buck shell is about the same size as the bullet from an army rifle. The standard 12-gauge 00-buck shell, with nine lead balls in a 1 1/8-ounce charge, can be rapid-fired from a six-shot pump gun so fast that the fifty-four slugs can do more damage at short range than a military submachine gun, which usually has a clip holding only twenty cartridges. And remember, the jacketed bullets do not have as much stopping power as the lead balls that mushroom when they hit.

A shell loaded with BB-shot will have some ninety pellets in a 1 7/8-ounce charge, or with No. 9 shot will have over 700 1/12-inch pellets in a 1¼-ounce charge. Shooting across a room with a properly modified shotgun even with such loads can hit a man in so many places simultaneously that they can knock him completely through any window behind him. At ten feet, the charge from a 12-gauge shotgun covers an area of four or five inches, with the virtual impact of a cannon shell.

Although the unwieldiness of a shotgun is a limitation, it is more than made up for in its ability to minimize the chances of the shooter's missing his target. You may never have to do more than point it at an intruder—staring down the 3/4-inch barrels of a double-barreled 12-gauge scatter-gun being held by an enraged homeowner takes the starch out of just about anybody.

Not all shotguns are suitable for home protection. A .410 or a 28-gauge gun does not fire a big enough charge, and is designed for shooting squirrels or rabbits, not people. The big 10-bore is more powerful than needed, and is designed as a goose gun for high-flying waterfowl. Most 16-gauge or even 20-gauge shotguns are adequate, although not as formidable as the 12-gauge.

Even some of these have limitations that should be

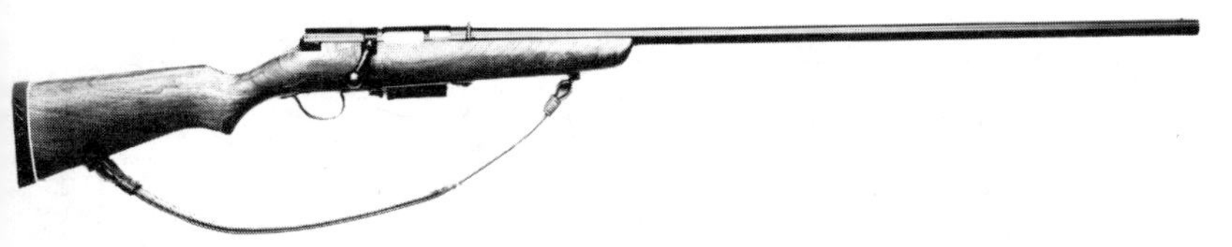

The powerful 10-gauge goose gun is too unwieldy for use as a house gun.

considered in buying a gun for home protection, such as the lightweights made to be less burdensome for hunters who carry them for long distances. A five-pound single-shot 12-gauge, for example, can have a terrific kick when fired. An eight-pound six-shot 12-gauge pump gun, equipped with a shoulder pad, absorbs most of the recoil itself in the weight, and the kick is barely noticeable.

Long barrels, up to 40 inches in length for turkey shooting, can also provide an excessive kick, but even worse, they are too cumbersome for the fast action a homeowner might need. Another disadvantage of having a long barrel on a house gun is that the barrel announces the user's presence if he comes around a corner with the gun pointed at the ready. If, instead, the user points the barrel up at the ceiling so he can peek around the corner, he may face trouble at which the ready-to-fire gun should be pointed.

Shotgun barrels available in sporting goods stores are *all* too long for the most effective use as a house gun. Most retailers are reluctant to sell a riot gun—a 12-gauge pump gun with an 18- or 20-inch barrel—to anyone but a law enforcement officer. The best that a homeowner can hope to buy over the counter is a slug gun, designed to shoot deer in states that prohibit hunting with rifles because their killing range is great enough to be dangerous even in sparsely populated areas.

The 12-gauge slug gun has a smoothbore barrel like any shotgun, but the single slug it fires has spirals cut into its diameter to make it spin as it comes out of the muzzle—just as a pistol or rifle bullet does. This slug is so heavy that it is effective only up to 100 yards or so. But the homeowner won't be shooting slugs, he will be shooting pellet shells, all grades of which can be fired from a slug gun. Most slugsters are multi-shot pump- or bolt-action guns (see below). The barrel length is usually 24 inches, fitted with sights like those on a rifle, with bolt-action models costing about half as much as pump guns.

The barrels on all other shotguns should be cut down to manageable lengths when used as house guns. Local laws covering sawed-off shotguns vary from state to state, and some of them do not even specify barrel lengths. Federal law, however, allows 18-inch barrel lengths for slug guns. In any case, it is best to have the modification job done by a gunsmith, who will know more about prevailing laws in the area than anybody who simply puts the gun in a vise and goes to work with a hacksaw. The gunsmith will also do a cleaner and safer job.

The stocks on most "sporting" guns should also be cut down when used for home protection. The pointed gun is often held at (and fired from) waist level, and the proper stock length should provide a straight reach from the inside of the elbow to the trigger (or triggers). Even many riot guns have stocks too long for this, as if the manufacturer figures that they will be fired only from the shoulder.

Cutting down the stock more than necessary, though, can mean legal trouble if the weapon can be considered a smoothbore pistol, capable of being carried as a concealed weapon. Gun collectors violate these laws every day with possession of antique arms such as duelling pistols (usually smoothbore as well as without sights so that the chance of duellists hitting each other was minimal), muskets, and eighteenth-century military arms. But such laws are on the books in many states, even if not enforced in all circumstances, and there is no good reason to rashly violate them.

Trap guns incorporate a lot of expensive features that are of no use at all to the homeowner.

Considering the way the homeowner will mutilate his shotgun to make it most effective as a house gun, there is not much point in buying a really good one just for home protection. A number of features that go into the cost of a good shotgun are meaningless.

The chokes in the barrels of sporting guns, for example, constrict the shot so that it stays together in a mass to make a tight (and more deadly) pattern when shooting game at relatively long ranges. Skeet and trap guns, except when used by real experts, are semi-choked to various degrees for a wider pattern in breaking the clay pigeon targets in the air. But the homeowner doesn't want any choke at all because he wants the shot to come out the muzzle to make as wide a pattern as possible at short range. The negative "advantage" of any choke system is eliminated when the barrels are shortened, as these constrictions don't even start until 3 or 4 inches from the original muzzle.

The ventilated rib down which the shooter sights along the top of the barrel is also an expensive shotgun feature the owner of a house gun can forget about. The ventilated rib is indeed of value to, say, the fortunate duck hunter who may fire several dozen shells within a few minutes; a shotgun fired that way can get pretty hot, and the air cooling off the matte-finished sighting rib minimizes heat wave distortion that occurs along the sighting plane. But a homeowner is not going to fire the gun enough times to get any barrel that hot, and it is unlikely that he will be so concerned with precision sighting that he'll be worrying about any heat distortion.

A house gun also does not need the durability factors built into the beautifully machined skeet and trap guns. These shotguns are sometimes fired more times in a single afternoon than a hunter will fire in a lifetime—and the householder will fire his shotgun only enough to know how to handle it (hopefully, in actual use, never).

Really fine shotguns like the Brownings belong in the field, not under a bed. A cheapie from Western Auto Supply will serve the householder as well as a handcrafted Belgian beauty with superposed over-and-under barrels and museum-worthy engraving. Even the trap gun made by Marlin, a company not renowned for top quality, costs about $235.

Likewise, the homeowner does not need a shotgun chambered for the new 3-inch magnum shells. The magnums can carry a bigger volume of shot as well as a bigger volume of powder, but they are for shooting at the type of game and at distances only a 10-gauge could handle before the magnum shells were popularized in 1973 and 1974.

Shotgun Actions

The lowest priced shotguns are the single-shot models made by companies such as Savage, Stevens, and Ithaca, or made under private label for discount stores and the mail order companies like Sears, Ward's, and Alden's. Singles are available all over the country in the $40 to $50 price range in all gauges (in most instances, the gauge size makes no difference in the price of a shotgun for any given make or model). Mechanically, single-barreled shotguns are quite simple and practically never get out of order.

They are safer to keep loaded in the house than the hammerless double-barreled shotguns, even if not as safe as a revolver with an empty chamber under the hammer, because the singles have a hammer that must be cocked manually before the gun can be fired by the trigger—a safety feature made necessary because the single-shot scattergun has a shell in the chamber at all times when kept loaded. Its biggest disadvantage is that the user gets only one chance—and he had better not miss with it.

Modern side-by-side double-barreled shotguns are almost as simple as the singles, with top-breaking breech action for loading and all, except that they are hammerless. Like automatic pistols, they are cocked and ready to fire if kept with shells in the chambers. They have a thumb-slide safety on top at the breech, but can nevertheless be dangerous to keep around the house loaded. The hammer springs may also be weakened by being kept under tension all the time.

Nothing else rivals the double in speed for getting off a second shot, including an automatic. Anybody fool enough to put two fingers in the trigger guard and pull both triggers simultaneously could indeed fire both barrels at the same time, although the resultant fire power would be like shooting off a small cannon and the recoil could be enough to knock the shooter flat on his back.

Not surprisingly, double-barreled models cost twice as much (and more) than single-barreled shotguns because, in effect, the buyer *does* get two guns with not only two barrels but also two firing chambers, two actions, and two triggers. Users who anticipate varying circumstances can load one side with one size of shot and the other side with heavier (or lighter) shot, or even include a shell loaded with tear gas in one of the barrels. The user can decide which barrel to shoot first—the barrel on the right-hand side with the forward trigger or the left-hand barrel with the trigger behind it.

Some double-barrels (the more expensive ones, with prices starting at $150) have only a single trigger, which is simply pulled twice to fire both barrels. In a gun meant for the same loads in both barrels, the firing order cannot be changed. Others have a thumb button with which the user can select the barrel he wants to shoot first. The advantages are moot. In all cases in which the user has different loads in the two barrels, he should handle the gun often enough so that he won't forget which side is which in the excitement of an emergency.

Superposed (over-and-under) double-barreled shotguns are made mostly for skeet and trap shooters who can afford to shoot up $50 or $75 worth of shells for an afternoon's amusement and who want the finest gunsmithing they can get. Superposed shotguns have everything a homeowner needs in a double-barrel (and more), but they do have the disadvantage of not looking as lethal as a side-by-side double. Finding a good over-and-under for less than $200 is not easy, and prices go all the way up to a couple of thousand dollars for some of the more exactingly crafted English and Belgian guns.

Multi-shot bolt-action shotguns are relatively cheap, mainly because there's little demand for them—they are too slow in rapid-fire situations. The shooter has to take his hand away from the trigger, lift the knobbed bolt handle and pull it back to eject the spent shell, shove it forward again to feed the next shell into the firing chamber, and lock down the bolt before he can get his hand back to the trigger again. Most bolt-action shotguns can be fired three times without reloading—with one shell in the chamber and two more in a spring-fed clip under the breech. They are available for $40 to $60, including short-barreled slug models.

Pump guns hold four, five, or even six shells in various makes and models. They are sometimes called "trombones" because of the way a shooter uses his left hand (for a right-handed shooter). To get a shell into the firing chamber, the left hand pulls back the wooden forearm under the barrel, which ejects the spent shell; sliding it forward feeds another shell into the firing chamber and cocks the gun automatically. Rapid-fire users like pump shotguns because, after a couple of charges have been fired, they can insert more shells into the bottom-loading port during a pause in the firing so that the tubular magazine under the barrel is full again.

Federal laws now prohibit the use of any repeater holding more than three shotgun shells when hunting migratory wildfowl (although there is no law against carrying as many loaded shotguns as you want). Most manufacturers "comply" with this law by inserting a plug into the pump gun's magazine so that it will take only two shells. Maybe they are unaware that the loose plug can be jiggled out quite easily by anybody who wants the magazine to hold a full complement of shells.

The standard safety on a pump gun is a trigger bolt that keeps the trigger from being pulled back. It has to be locked manually by pushing it to the right when the user wants the safety on. When ready to fire, the user depresses the locking bolt located immediately in front of the trigger. Not counting the fancy models, pump guns cost from $75 to $125.

Many manufacturers also make automatic shotguns now, which have gas-operated actions working the same way as the actions in automatic pistols. They have all the same advantages and limitations as the handgun automatics (see above). They may be okay in the field, but not in the house. Standard three- to five-shot automatic shotguns cost from $175 to $200.

Rifles

No rifle ever made, except maybe an old-time muzzle loader loaded with bird shot, is really suitable as a household defense weapon. Rifles are meant for long-distance shooting, not short-range firing, and can accidentally kill anybody in the line of fire within a mile or more. Most of them do not have the stopping power of a heavy handgun at short range, much less that of a shotgun. Shooting somebody with a .22 is more likely just to get him mad than to kill him, and in most circumstances could best be used as a weapon if turned around and used as a club. Even a deer rifle has to hit a vital spot to knock a man down.

The best of the lot would be a lever-action saddle gun. These short-barreled carbines use relatively low-velocity .30-30 cartridges, and the repeat action is much faster than with a bolt-action rifle (which is meant for shooters who have plenty of time to aim). The most popular is the seven-shot Winchester "Western" Model 94, with a tubular magazine under the 20-inch barrel, selling for around $80. Because it has a manually operated hammer, it is also safer when kept fully loaded around the house than the hammerless slide-action and gas-operated automatic rifles. The venerable Model 94 has been unchanged in many, many years, and Marlin's copy, the Model 336T at only about $35 more, incorporates a good many modern improvements.

Safety

The first thing any firearms buyer absolutely must do is to become familiar with the weapon if he is ever to use it effectively. Some owners who "keep a gun in the house" don't even know how to load it, how to take the safety off, or how to cock it, and don't know either what its destructive capabilities can be or what its limitations are.

Even if you do know something about guns in general, you should familiarize yourself with the gun you have chosen to buy.

Being a veteran of the Armed Forces does not mean that an ex-serviceman is proficient in the use of firearms for self-defense. Most military training is concentrated on the use of automatic and semi-automatic weapons at long range. A good many local gun clubs have instructors, but their interest, too, is usually in long-range shooting.

A gun buyer should find a shooting range (most police departments have one) and shoot up at least one box of

The only rifle of much use for home protection is a short-barreled saddle gun.

ammunition to get acquainted with the weapon. If he is inexperienced with it, he should get somebody who can show him how it works. Even after this initial practice, the gun should be fired a few times every once in awhile to keep the owner in practice. Even dry-firing the gun, with spent shells in the chambers so as not to damage the firing pin, is better than nothing.

The gun should be cleaned after each use, and every month or so it should be cleaned whether used or not. Cleaning kits, which require only a modest investment, are made for specific sizes and kinds of guns, and they do a much better job than just running a rag through the barrel and cylinder chambers.

No gun is of much use in an emergency unless it is kept loaded (even if you only keep an empty chamber under the hammer for the sake of some safety). It should be kept within easy reach of the user, but out of the reach of children. The latter is easier said than done; many gun owners use a key-operated trigger lock, such as the one made by the Master Lock Company, that completely shrouds the trigger guard so that the trigger cannot be even touched.

Anybody handed a gun for inspection should always assume that it's loaded, regardless of what the owner says. This is easy to check on a double-barreled shotgun by breaking it open to see if the barrels are indeed clear, or on a revolver by opening the gun to see if there's anything in

A locked trigger shroud can prevent accidents, although unloading the gun would be safer yet (a revolver can be hammer-fired without touching the trigger).

the cylinder. On slide-, bolt-, automatic- or lever-actions, the action has to be worked twice to make sure that the gun is empty (working it only once may not eject a cartridge, but indeed may feed a live cartridge into the firing chamber from the magazine). No gun, loaded or not, should ever be pointed at a living target, because, as noted before, there's only one reason to ever point a gun at anybody.

Membership in the National Rifle Association doesn't

do a gun owner much good unless he is actively interested in guns beyond the mere ownership of a house gun. The NRA has over a million members, a good many of whom insist on the unrestricted ownership of guns "to develop marksmanship as an aid to national defense."

Membership in the NRA costs ten dollars a year, including the excellent monthly magazine. Applicants have to be recommended by a member in good standing (dues paid up), by a commissioned officer in the United States Armed Services, or by "a public official" (which could presumably include a number of rascals). National headquarters are at 1600 Rhode Island Avenue, Washington, D.C., 20036.

One final word: one danger in owning a house gun is inadvertently walking in on a burglar who has already ransacked the house and thereupon uses the householder's own gun on him. The problem, often cited as a reason for not keeping a gun in the house, is easily solved by unloading the gun and taking the cartridges with you when you leave the house. If the burglar then points your own gun at you when you come home, you'll know that it's unloaded and you can feel free to clout him one.

Unloading a gun is no more touble than locking the door, and it should be considered a routine part of the locking-up process whenever leaving the house. Keep only enough ammunition in the house to keep the gun fully loaded, with the spare ammunition at the office or at a neighbor's house.

There's no good reason to be afraid of keeping a gun in the house for self-protection. Let the *burglars* be afraid.

5

Watchdogs

"When I know there's a male watchdog on guard, I take along a female dog in heat to distract him, so I can work undisturbed while they're preoccupied. The D.A. looked down on me more for being a kind of canine pimp than for being a burglar"—Convict, Sing Sing Prison, Ossining, New York

A dog guarding the premises isn't worried about his owner's property. His only concern is repelling any encroachment on *his* territory, and he will attack another male dog even more viciously than he will attack a human trespasser. If a dog is to be an effective watchdog, he must be allowed to develop his territorial instinct. A dog who is allowed to sprinkle the fireplug and bushes around his house is a better watchdog because he is letting the world know that he has staked out his claim and marked it with his scent.

When a homeowner gets a new dog, the dog is unlikely to be much good as a watchdog until he has been made to feel at home and to develop a sense of territorial ownership. When he hears strange noises, at least for the first few days, he is more apt to cower than to bark; for all he knows, until he feels that sense of proprietorship, he might be on some other dog's turf. To a dog, the command "Sic 'em!" means "he is a threat to your territory."

Acuteness of Senses

A dog's senses of smell and hearing are much more acute than a human's. A bloodhound's sense of smell is so dependable that it is sometimes admissible as evidence in court, as in picking out a suspect from a lineup. These remarkable dogs can track a person by picking up the scent left on objects the person barely has brushed against.

Even the most spoiled poodle can distinguish between different odors that would baffle a human. If the wind is blowing toward him, a dog can detect another animal (man included) as far as a couple of blocks away. Dogs can operate in almost the same way as a lie detector in discerning the almost imperceptible perspiration exuded by people under stress, which would include a burglar at work. "Stress perspiration" contains an ingredient that irritates dogs, which is why they are prone to attack anybody who is afraid of dogs, the mailman included. Professional dog handlers who are not afraid of dogs can walk through a kennel full of strange dogs with nary a yap forthcoming; the same dogs would all but tear the place apart if somebody scared of dogs should walk in. This instinctive reaction to fear is an inheritance from the wild dogs in the dim past who depended on fear to help them select the prey on which they lived.

Dogs can also hear sounds that are inaudible to the human ear, with a far greater range of perception in both higher and lower sound frequencies. One of their most remarkable traits is the capability of distinguishing between the sounds of different footsteps. They either ignore or welcome the footsteps of members of the family, but growl or bark at the approach of strangers before they can either see or smell them. Because of these inherent abilities, naturalists agree that the most successful predator in nature is not the lion or tiger, but the wild dog.

Most burglars will avoid a house with a "Beware of the Dog" sign on the gate. Considering what he has at stake, the average burglar will never take a chance if he has any alternative—and there are plenty of alternatives in most neighborhoods. Some people buy a "Beware of the Dog" sign even if they don't actually have a dog. In such cases, they should also buy a dog's feeding dish or even a doghouse to make the appearances more convincing.

Watchdogs should be trained never to accept food from a stranger. Tossing a dog a chunk of hamburger loaded with knock-out drops is all too easy. Some burglars do not hesitate to poison a dog with lethal toxics—the penalty for killing a dog is no more than for the destruction of any other kind of personal property.

Aggressive Dogs

The homeowner who keeps a vicious dog stands the same legal risks as a man who keeps a loaded gun in the house, and more if the dog bites any innocent people. If the dog is trained as an attack dog, such as the killer dogs in the army's K-9 Corps, the owner can be charged with using deadly force if somebody is injured (see previous chapter). Dogs that have had attack training cannot be untrained, and most of the K-9 dogs had to be destroyed instead of being deactivated and returned to their former private owners.

An attack dog cannot be trusted even with members of the family, let alone visitors. Doberman pinschers and German shepherd police dogs can be raised as house pets if properly trained from puppyhood, but once their natural instincts for aggression have been exploited with attack training they are too dangerous to keep around the average house. Dobermans, in fact, have a nasty inclination toward lying in wait for somebody to come in, not making a sound until the intruder is within reach, and then making straight for the throat. If the "intruder" happens to be Uncle Louie coming home late, an attack-trained Doberman will tear out his jugular as readily as anybody else's.

"Nice Shep" can kill an intruder if trained to do so, and a watchdog around the house is better trained if taught to protect his property/territory just by barking. This fellow only looks like he's smiling; those hind legs are ready to spring to the attack.

Many strains of Dobermans have been deliberately bred to make the most of their aggressive instincts, with successive generations of the most vicious males being mated with only the most vicious females, to produce supervicious animals that need little or no training to become attack dogs. The child who puts his hand out to that kind of "nice puppy" is all too likely to lose an arm. The Doberman is bred to have short ears and hair to make him hard to grab, powerful hind legs to give him speed in turning, and a powerful bite that enables him to exert 600 pounds of pressure per square inch.

The only other dog that can match the Doberman for raw courage is the dachshund. He has been bred to hunt badgers—short legs to get down into burrows, long ears to keep the dirt out, a big chest for lasting power in a fight, and a long snout for biting. The badger is a very tough customer indeed, and dachshunds have been bred to fight them to the death. A dachshund will rashly attack even Great Danes and German shepherds if they wander into his yard, and in most cases he will run them off, too.

Like the Doberman, the dachshund doesn't just bite; when he sinks his fangs into somebody and gets a good hold, he shakes his head violently from side to side so that he rips and tears. He is generally a one-man dog, attaching himself

Dobermans are specifically bred to be fighting machines . . . powerful hind legs for speed in attack, short ears and hair to make grabbing difficult, and a long snout with long fangs—plus an inclination to use them.

to one member of the family and tolerating the rest without really trusting them.

Training

Feisty as he is, the dachshund has never been trained as an attack dog because he just doesn't have enough weight to knock a man down. The Doberman, which does have the heft, is the official combat dog of the U.S. Marine Corps. The Air Force, however, uses German shepherds because their dogs are used primarily as sentries, not killers. The Air Force is the biggest user of dogs of all the armed services.

The Air Force takes dogs between one and three years of age. Dogs less than a year old are too much trouble to train, and dogs more than three years old are too set in their lifestyles. Females can be used for sentry duty only if they are spayed, because a female, when she is in heat, is too undependable.

Military dogs are trained and cared for by one man, and one man only. They do not come in contact with anybody else except in an attack situation. The result is remarkable efficiency, but only for military duty. A house dog has to get along with everybody in the family, not just one person. If Junior takes him for his walk, the dog has to respond as readily to the youngster's command to "Heel!" as if the head of the household were walking the dog himself.

A well-fed dog decently cared for needs very little training to be a good watchdog. When treated as a house pet, and taught to know the simple limits of what he is allowed to do, the dog's natural instincts are more dependable than any formal training, especially when he considers himself a member of the family.

Sending a dog to an obedience school, at an average cost of a couple of hundred dollars, can teach a dog a lot of tricks, but it won't make him a better watchdog than his innate character will make him. A dog has a strong desire to please his owner, and the only problem the owner has is in establishing communication so that the dog knows what's expected of him.

Certain breeds of dogs do have innate characteristics that make them particularly good watchdogs. Besides Dobermans and German shepherd police dogs, burglars are also afraid of being attacked by Airedales, Russian wolfhounds, chows, huskies, mastiffs, schnauzers, spitz, bloodhounds, dalmatians, labradors, and weimaraners.

Many police chiefs are convinced that the best kind of a watchdog to have around the house as a deterrent to burglars is a "yapper." These small, nervous dogs are too scared of everything to try to bite anybody, but they will hide under a davenport and continue their shrill barking as long as there's anybody in the house the dog thinks shouldn't belong there. Yappers include Chihuahuas, Pekingese, Pomeranians, pugs, and toy poodles. None of them could win a fight with a healthy kitten, or would even try. But their high-pitched yapping can wake up an entire neighborhood.

Another much-favored watchdog is the ferocious-looking English bulldog. With long, dripping fangs sticking up from his lower jaw and a pushed-in face that makes him look as if he were a veteran of countless battles, his very appearance is enough to strike terror to the heart of a trespasser, even if the dog never so much as growls. And he seldom does; this chap seems to have forgotten that his ancestors were originally bred for bull baiting and for pit fighting, and is now one of the most affectionate dogs in the whole canine family.

Most big dogs, which need a lot of exercise, should be kept only in homes where there is a big yard for them to roam around in. One exception would be the Saint Bernard, who is quite comfortable in relatively small areas because he is so big that he doesn't romp around as much as smaller dogs. The collie, television's Lassie notwithstanding, does not have all that much intelligence and is not as good a watchdog as many other breeds that do not require as much care.

Many medium-sized dogs need a great deal of romping room, including such breeds as beagles, wire-haired terriers, Scotties, boxers, and bigger poodles, all of which have a lot of energy. Mongrels are not usually as high-strung as the purebreds, they are easier to take care of, and they are often of hardier stock. The main trouble with a common mutt is that the owner never knows what the puppy will grow up to be; whereas a registered dog's lineage is known, and his adult characteristics are more predictable.

Pedigreed Dogs

The purebred is much less pure than is commonly supposed. The civilizations of Asia Minor developed the coursing or "chasing" dogs, and the tracking and bird dogs were developed in Europe. But the development of distinct types, as recognized today, started only about 100 years ago, with the advent of dog shows. The first dog show was held in England in 1859, with American dog shows beginning around 1870, but most breeds did not become "fixed" until after 1900. The Doberman pinscher, today one of the purest of purebreds, was developed as a personal project of Louis Doberman of Apolda, Germany, around 1890.

The American Kennel Club now recognizes 120 breeds, with some split into varieties, bringing the total to 135 breeds on which stud books are kept. Although the AKC registers more than a million purebreds a year, the exact number of purebreds in the U.S. is unknown because more than half of the dogs eligible for registration are never registered. The total U.S. dog population is estimated at 30 million animals, including some 15 million mixed breeds.

Just because a dog "has papers" doesn't necessarily mean that he is a better dog. It only means that he has been registered, for better or worse; his entire line of ancestors may have been thoroughly rotten dogs, and their registration is nothing more than a record in the stud book.

Most hunting dogs have been bred to be such specialists that their sense of aggression has been blunted too much to make them good watchdogs. All three types of sporting dogs—pointers, flushing breeds, and retrievers— have been bred to be better and better at less and less until they are truly excellent at only a narrowly defined job. Some water spaniels, for example, are unmatched for retrieving waterfowl from water, but will retrieve it *only* if it has fallen into water.

Hounds, as distinct from sporting dogs, are trackers specializing either in following by scent or hunting by sight. The latter include greyhounds and whippets built for speed, an aptitude of not much use to a homeowner whose primary concern in getting a watchdog should be to get an alert dog with a loud bark.

Ownership Costs

A good watchdog can cost anywhere from the cost of a dog tag at the city dog pound to $1,000 or more for a fully trained weimaraner or Doberman. Owning a dog involves more than just the cost of the dog food. He will need a collar and leash, feeding dishes, and toys. He will have to get rabies and distemper shots, worming treatment, and good veterinary care from time to time if he eats something that makes him sick, gets scratched up by a neighborhood cat, or gets knocked down by a passing automobile.

If the owner acquires the dog as a puppy, there's likely to be considerable damage around the house before the dog is properly trained. And mainly, the dog will require a lot of the owner's time. Once the dog has established a routine for taking his walks, the owner must adapt his own lifestyle to match that routine. If the owner doesn't have other members of the family who can take care of the dog, there's no such thing as stopping off at a friendly saloon for a couple of hours after work if the dog is accustomed to taking his evening walk at 5:30.

When the dog is in the habit of taking his final walk of the day at 10 P.M., the owner will have to forget about going to late movies or partying until the wee hours. When it's time for the dog to take his walk at 8 A.M., there's no late sleeping for the owner on a morning after, either, including Saturdays and Sundays. These are some of the reasons why single people seldom own dogs.

Another reason is that dogs are jealous. If the dog's owner invites in a date for a nightcap, a whining dog scratching at the door can put quite a damper on the proceedings. Many dogs actively resent a new spouse moving into the house, and they do not readily approve of the arrival of babies in the family.

Having a watchdog in the house should never be considered as a substitute for good locks and other sensible security measures. But for a sense of security, there's nothing else quite in the same class as having a four-legged burglar alarm on the premises.

6

Burglar Alarms

"Not all crooks wear masks. Some of the worst thieves in the burglar industry include the suede-shoe guys who sell and install grossly oversold and overpriced burglar alarm systems"—Convict, Leavenworth Penitentiary, Leavenworth, Kansas.

"How would you like to wake up in the middle of the night and find a rapist crawling in through your bedroom window?" said the garishly printed circular distributed door to door in New York. "Doesn't it chill your blood to think that you might find your children murdered in their beds? Our revolutionary new burglar alarm gives you absolute and complete protection, and is approved by the FBI for 100 percent dependability because it is hooked up directly to your local police station. You will have officers on the premises within minutes after an intruder just *tries* to break in. Call us NOW *before* tragedy strikes . . . for a free demonstration at no obligation."

Scare tactics like these, used in under-the-door literature in middle-class neighborhoods throughout the city, generated thousands of inquiries in the early 1970s. The salesmen who showed up gave the impression that they were connected with "the Crime Commission" and were part of a public-spirited crusade to cut down the alarming increase in crime. They were armed with plenty of shocking statistics, including an official 280-page copy of *Crime in the United States*, published annually by the Federal Bureau of Investigation, U.S. Department of Justice, Washington, D.C.

The salesmen also carried portfolios of gory photographs showing ransacked houses and mutilated bodies. The "demonstration" was actually a demonstration in how easy it would be for a burglar to break in . . . including carding a spring latch, jimmying a deadbolt, butter-knifing a window lock, and even picking locks. Some of these boys were so good at their demonstrations that police subsequently tried to find them on the supposition that they were, in fact, bona-fide burglars who had found a safer line of endeavor.

By the time these "salesmen" were through with their sales pitch, they practically had the terrified householders begging to buy their burglar alarms for the "pennies a day" in purported costs. Those "pennies a day" added up to an iron-clad sales contract usually ranging between $500 and $1,000. Some contracts topped $3,000. In some cases, signatures were obtained fraudulently through purported "receipts" for equipment left for what buyers thought were free trials.

Bess Myerson, in charge of Consumer Affairs for New York, got so riled in finding that one of the burglar alarm companies was charging an average of $750, for maybe $75 worth of equipment, that she instigated a full-scale investigation. Ms. Myerson found out almost immediately that there was no connecting alarm to the police station as so blatantly advertised; if a woman heard her bedroom window being raised and sounded the alarm, nobody could hear the alarm but her.

Bess subpoenaed the records, talked with just about everybody who had been sucked in on the deal, and succeeded in getting many thousands of dollars back in refunds for the deluded. Any reference to the Crime Commission, much less to FBI approval, had also been fictional. That impressive FBI report is available to anybody who will send $2.85 to the Superintendent of Documents, Government Printing Office, Washington, D.C., 20402. Some of the people in the gory photographs used in the sales pitch turned out to be actors with catsup poured on them. The only thing revolutionary about the "new" burglar alarm systems—actually quite ordinary gear that has been used in commercial installations for many years—was in the new sales approach for residential use.

The New Yorkers who got their money back were lucky in having somebody like Bess Myerson on the job. The same ploy, with modifications and improvements, is being used all over the country. If you, as an average householder, have not yet been solicited to buy a burglar alarm system, you probably will be soon.

One of the worst results of using scare tactics to sell burglar alarms is that they unnecessarily spook too many people. The overwhelming majority of burglars are not rapists and murderers, they are sneak thieves. Compare the nearly 3 million burglaries committed in the U.S. last year, to the 18,000 or so murders. Some 75 percent of the murders are committed by relatives or acquaintances, leaving fewer than 5,000 "real" murders—which means that the odds are some 600 to 1 that a burglar is ever willing to shoot anybody. If somebody has rape or murder in mind, security systems won't keep him out because the victim usually knows him and *lets* him in.

The only thing a burglar alarm is good for is to scare away burglars. To sell it on any other basis is a disservice to the industry.

The burglar alarm industry is very big business indeed. The Chicago classified phone book, for example, has over nine full pages of listings under Burglar Alarms in the Yellow Pages. In 1974, the International Security Conference, which all but filled up the Conrad Hilton Hotel in Chicago, had 198 booths in the exhibit halls. Most of these manufacturers and distributors are conscientious people trying to do their best in a tough job. But it is to the advantage of the industry to see to it that the fly-by-nights and fast-buck operators who give everybody else a bad name are exposed for what they are and put out of business.

Even householders who deal with reputable firms can sometimes find that they have been victimized by crooks who have infiltrated such companies. A larcenous installer of burglar alarms, or a crooked repairman, can make even the best burglar alarm system worse than worthless—by selling the names, addresses, details of installation, and the house plans to a ring of burglars.

When a burglar knows what to watch out for, where it is, and how to disarm it, the householder would have been better off just to get a good watchdog. Some manufacturers recognize this weak link in their security systems by building in "owner adjustment panels." The householder can use this panel to change the entry codings after the installer leaves, so that only the householder himself knows how to get into the house without setting off the burglar alarm.

Components

The basic components in most good burglar alarm systems have three functions: (1) a sensor, which detects intrusion; (2) a reporting device, such as a loud siren or a system that notifies an outside agency; and (3) a control system, which arms and disarms the burglar alarm for the owner's convenience.

Sensors can be break-sensitive tapes like the "silver decorating lines" used near the edges of doors and windows in many stores, weight-sensitive mats placed under carpeting or rugs at entrances, physical switches that break electrical contacts when a door or window is opened,

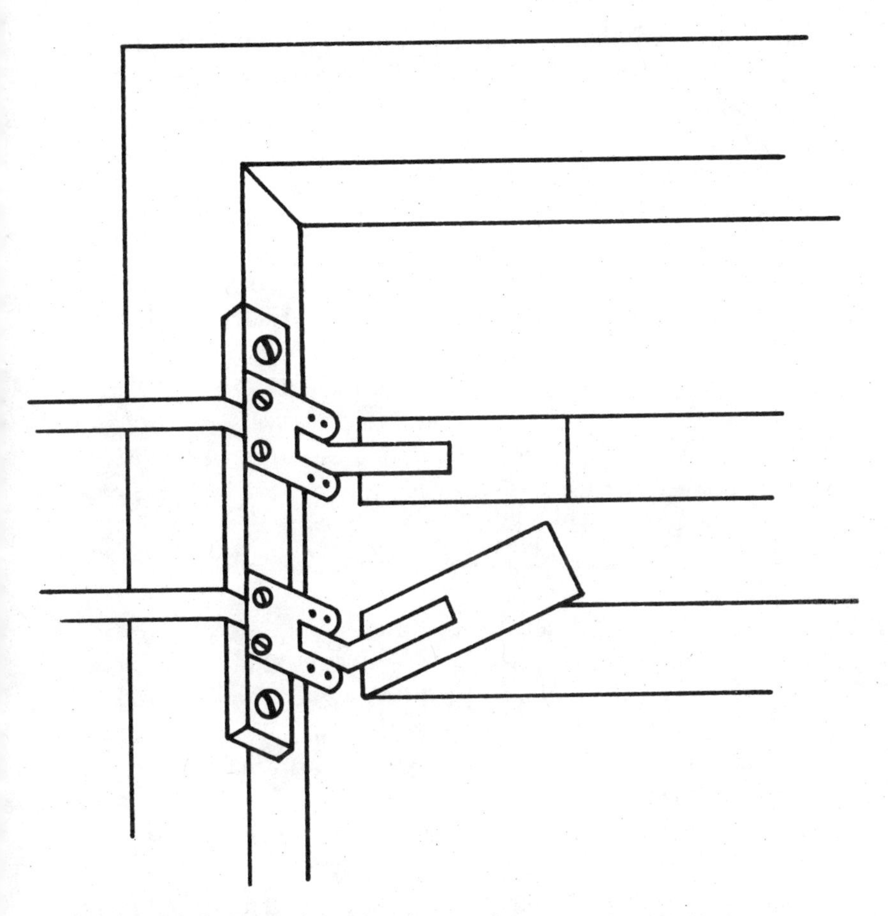

Insulated foil mounted on glass is one of the most dependable intrusion detectors for a burglar alarm.

magnetic cleats, electric eyes, ultrasonic and microwave detectors, or infrared light beams.

Most burglar alarms utilize interior space systems, not only because no burglary has been committed until after the miscreant has broken in, but also because they do not have to be weather-proofed. But perimeter systems, guarding accesses to the house, are also growing in use with the simple idea of chasing the burglar away *before* he gets a chance to try getting inside. Not one sneak thief out of a thousand will persist in trying to break in once an alarm has sounded.

Silent alarm systems, in which an outside agency or police station gets the notification, give the police a better chance to apprehend the criminal. This is fine if the homeowner is more interested in his civic responsibilities than in his own property, but chasing burglars away is all that most people are immediately concerned about.

A clanging burglar alarm, especially when it also activates flashing lights, is enough to discourage almost any intruder. A siren is even more effective than a bell because most people associate the warbling tone with the sound of a police emergency vehicle.

Basic Bugs

A number of detectives interviewed at the International Security Conference in Chicago cited as unnecessary the number of sensors a homeowner would need to cover *all* entries to the house. If he has not only a front entrance and a back door, but also a patio door, an entry from the garage through a breezeway, a basement door, and numerous windows all over the main floor, seond floor, and basement, his investment in sensor devices could be considerable. All would have to be kept in maintenance, too. Instead, said these experienced detectives, put a single burglar alarm at the bathroom door.

The first thing a burglar does is peek in all the rooms to make sure that nobody's in the house. He then opens exits front and back so that he can get out fast in case somebody comes home unexpectedly. By that time he is so nervous that he has to make a visit to the john. An alarm mat placed on the floor at the door is notably effective, and it can be disarmed when the family is home simply by picking it up and hanging it on a nail.

Disarming a burglar alarm, so that members of the family can come in and go out without arousing the neighborhood, can involve complicated devices incorporating keys and push-button codes, or can be simple hidden switches that deactivate the alarm system. The latter have severe limitations because, if they are outside, burglars can find them. One alternative is to have dummy switches in the most obvious "hiding" places, which turn *on* all the lights and sirens when touched.

Fakes

Many of the best burglar alarm locks (shunt locks) utilize tubular keys because that kind of lock, with its pins extending from the back toward the keyhole, is so hard to pick (see Chapter 1). This type of shunt lock, even when not connected to an alarm but simply put into the jamb alongside the front door, is enough to discourage most burglars merely by its presence. To a burglar who knows what it is, it means trouble; he either will go elsewhere without further ado or will drive himself to distraction trying to disconnect the nonexistent wires. Any burglar who doesn't know enough to recognize a tubular-keyed burglar alarm shunt lock is unlikely to know enough to jimmy through a one-inch deadbolt, either.

For a similar purpose, many people acquire window decals that say that the house is protected by an electronic security system. The use of fake decals, without having an actual burglar alarm system, is most effective when the decals do not carry the name of the alarm manufacturer. Too many burglars know too much about the different makes of burglar alarms, and by looking over the house they can often tell whether or not the make noted on the decal has actually been installed on that particular house. As far as that's concerned, even the decals on houses that do have alarm systems should not include the name of the manufacturer because it just might give a knowledgeable burglar more information than the householder would want him to have about the type of system that would have to be disarmed.

Another fake sometimes used is the burglar alarm call box, most often a cast-iron box painted red and mounted high up on an outside wall under the eaves. These are often left by a former owner of the house who did indeed have a burglar alarm system but who found that the call box was too hard to remove when he moved out and took the rest of the security system with him. If the warning decals were also left on the doors, the new householder has almost as much protection against "burglars of opportunity" as the buyer of the real equipment had, in deterring an attempt at intrusion.

Limitations

One disadvantage of any recognizable burglar alarm system is that it announces to the burglar, "There's something worth protecting inside." A burglar alarm system may scare off the amateurs, but it can attract professionals.

Professional burglars have better things to do with their time and talents than rummaging through somebody's dresser drawers. On odds, there's more profit in commercial and industrial hits. But when burglars see signs of an installation they can recognize as a $1,000 or $1,500 burglar alarm system on a house, they can be fairly sure that there's something valuable inside. If they recognize it as an old-fashioned hard-wire system that operates off regular house current, all they have to do is wait until nobody's home and then cut the power line to disarm it.

Another disadvantage of any burglar alarm system is that it complicates the owner's life. All members of the family have to learn how to live with the system, as well as to learn the sometimes complicated process of operating it. A number of people have accidentally set off so many false alarms that they eventually disconnect the system in disgust, giving it up as a bad experiment.

The buyer of a burglar alarm system should bear in mind that its primary purpose is to protect the house from burglars when the family is away. (By definition, a burglar is a sneak thief who will attempt entry only when nobody's home.) However, there is very little sense in installing an expensive burglar alarm system to protect household goods of limited value. A great many $800 to $1,200 burglar alarm systems have been installed in homes where there is nothing inside worth taking beyond a five-year-old television set and a portable typewriter—unless you count the kitchen toaster and electric clock—and where everything is covered by insurance, besides.

A number of burglar alarm systems are sold to householders with the idea of letting them sleep more soundly with the knowledge that the house is wired up with an alarm to alert them when somebody's trying to get in. If flashing lights and a loud siren scare away a surprised burglar who thought nobody was home, all well and good. But too many burglar alarms sold on the premise of "awareness security" are simply wired to a bedroom alarm, and the burglar does not realize that he has triggered an alarm sensor when he enters. Unless the householder is willing to pick up a gun and kill the intruder, what's he going to do about it? From the standpoint of his own personal safety, the householder is far better off trying to scare a burglar away than trying to catch him.

Automatic Telephone Dialers

Many silent alarms incorporate automatic telephone dialers to contact remote communications centers or the police station. Automatic dialers can indeed be activated by burglar alarm sensors, but no faster than a householder can pick up the phone himself and dial 911, which is rapidly becoming the universal emergency number for police departments. In many ways, dialing 911 is better.

Dialing that number automatically locks in the calling number at the phone company so that police can trace the call immediately even if the phone was hung up before the message was completed. Also in its favor is the lack of accidental false alarms on 911, which happen so frequently with automatic dialers. A gust of wind rattling the windows, a power surge in electric house current, a low-flying plane, a short circuit, or any number of acts of carelessness on the part of the householder can bring the police on the run. Police departments all over the country are plagued with countless false alarms during every thunderstorm. This false alarm situation is so serious that the authorities in several parts of the country are considering legislation to make automatic dialers illegal.

Most automatic telephone dialing systems are far from foolproof as real protection. Many manufacturers recognize the problem of false alarms by building in an abort button on the dialer, which gives the householder up to thirty seconds to cancel a false alarm before the message goes out over the telephone line. However, even this safeguard has its drawbacks. If a burglar uses, say, an automobile bumper jack held horizontally at the height of the door lock, he can spread the door jambs far enough apart to pull an average deadbolt out of the receiver so that he can push the door open in five to ten seconds. Even if the door sensor activates the telephone dialer at the first pump of the jack handle, the burglar has plenty of time to get to the dialer and press the abort button.

Other dialers, in an attempt at eliminating as many false alarms as possible, have a twenty- to thirty-second time delay before the dialer actually starts making the call—which gives a burglar time to get in and simply lift the receiver off any extension phone in the house to defeat the dialer. The burglar can also block the average automatic dialer from making the emergency call simply by calling the

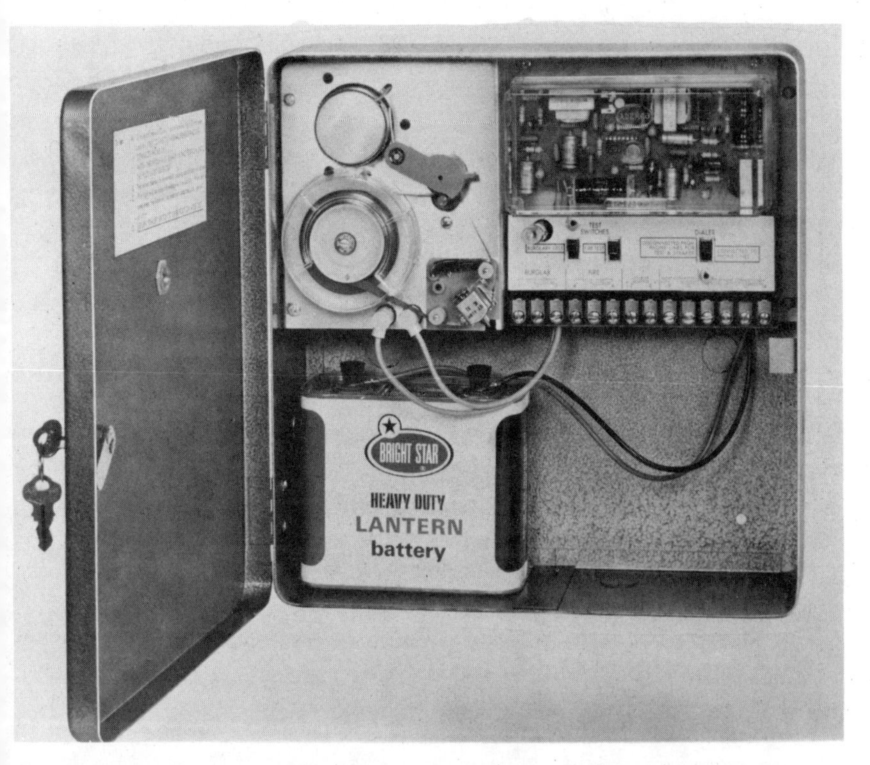

Automatic telephone dialers can dial the police whether the homeowner is home or not. But they can get out of order if not tested frequently and automatically phone in so many false alarms that their use is now illegal in some localities.

house from an outside phone to tie up the phone line.

In many cases, a burglar won't have to be so devious. A number of automatic dialers are no more than ordinary tape players. Their electromechanical solenoids depend on large amounts of current and break down easily; when left idle for months (hopefully, forever), the motors sometimes won't start, or the magnetic tape in ordinarily unreliable casettes will wrap around the drive mechanism, or the pressure rollers will develop flat spots from resting so long in one position that the tape won't play.

Good telephone dialers have standby power packs in case the house current fails and have two or more channels for different messages in different circumstances.

The telephone company and FCC rules tightly limit the speed at which dialers can transmit their pulses; when the pulses are too fast or too slow, the dialer is likely to misdial or not even dial at all. This is a far more common hazard than might be supposed. Many motors used on ordinary tape players vary in speed depending on the voltage. Still others might dial the pizza carry-out instead of the police station because of a poor alignment between the tape and the magnetic head or too much play in the capstan bushing.

A notable exception to dialers with these problems is the unit made by Alarm Devices Manufacturing Company (Ademco), Syosset, Long Island, New York. Ademco uses a motor that runs at constant speed regardless of different voltage outputs, with built-in speed governors, and uses a special continuous-reel cartridge designed by the company to minimize bind-up and drag. The Ademco dialer also has a precision-molded tape guide for magnetic alignment, a capstan with ball bearings instead of an ordinary bushing, and elastomeric rollers that "store" under minimal pressure.

The problem for the consumer: Ademco sells only to installers, and not to the homeowner. The company's automatic dialer may cost an installer only $125, but it might cost the user anywhere from $200 to $1,000 by the time it is programmed and hooked up to sensors, depending on the installer as much as on how much other equipment is installed with it.

Central Stations

Some of the best dialing equipment is used by the big security service companies that maintain remote communications centers of their own. Companies such as A.D.T. (American District Telegraph Company), along with divisions of Wells-Fargo, Burns, Norman, and Westinghouse, maintain twenty-four-hour security offices tied into their clients' burglar alarm systems. Many such installations utilize direct lines to the company's central station rather than using the regular telephone line. The practice of using mileage charges for direct lines is being phased out, and the phone company now charges a flat rate in most parts of the country, usually $50 to $75 a year for a residence in a metropolitan area. Westinghouse does not use tape at all for emergency messages; the householder's control unit sends a code signal, which is then decoded on a printout at the central station for indicated action.

Although all security service companies do a lot more than just monitor burglar alarms, Westinghouse is notable among them for the completeness of its service. Besides intrusion sensors, their system can also monitor fire and smoke detectors, water pressure lines, freezing pipes, furnace and boiler controls, flooding basements, emergency ambulance needs, and anything else that can activate a switch of any kind. It is also one of the most expensive systems; sold outright to the homeowner, a Westinghouse security system will cost the average buyer in the neighborhood of $2,000.

A.D.T., the biggest and best-known company in the security service field, also now sells their residential installations outright (although most of their commercial and industrial security systems are still leased). The average installation costs around $1,200 for burglary and fire protection, plus about $65 to $75 a year for service. Wells-Fargo is about the only big central-station security company still operating on a lease basis for homeowners, with an average $1,000 installation fee and around $800 a year for monitoring and maintenance.

Actually, individual installations will vary in cost because they are all custom-built depending on the size, floor plan, age, construction, location, and varying requirements of each property to be protected. The advantage of contact with a private twenty-four-hour central station is that police departments are not equipped to handle the variety of emergency calls that an automatic dialer might send when wired into a complete security system.

The sensors such companies use cover the whole range of residential security, from physical switches at doors and windows to photoelectric eyes, infrared beams, and ultrasonic and microwave devices. But "The Big Three" all stick to hard-wire systems, as the most thoroughly proven engineering.

The Protective Services Division of Norman Industries is somewhat unique in the central station field in that they use wireless sensors at points of entry for the burglar alarm. The door and window sensors resemble ordinary physical or magnetic switches, but they are not connected to the alarm control by wire. Instead, each sensor has a built-in high-frequency transmitter, which sends a radio signal to the receiver in the alarm control in an emergency situation.

The advantage of this wireless system is that a lot of labor is eliminated in the installation process; Norman can protect the average house for less than $1,000 on an outright sale.

Norman's system is unsurpassed for fire and smoke detection, but it has limitations as a burglar alarm. The main problem is that the system is not supervised; i.e., the lack of pilot lights on the master control means that when the householder goes to arm the system, the only way he would know if a door or a window had inadvertently been left open would be by the alarm going off. Another trouble with a radio-controlled system is that the alarm can be activated accidentally by other stray radio waves in the neighborhood such as garage door openers, ham radios, automobile telephones, or walkie-talkies. If the house is in the flight pattern of a nearby airport, false alarms can also be set off by pilots communicating with their control tower.

Some of Norman Industries' over-enthusiastic salesmen have been known to make claims that do not stand up under close examination. The company has had its share of lawsuits. The parent company jumps around in a lot of businesses, from television to space-age electronics, and its priorities shift constantly. Whether or not they will be able to put much of a dent in A.D.T.'s domination of the security service field remains to be seen.

Radio Sensors

Hardware installers (with no central station service) already include emulators of Norman's radio frequency sensors. The most noteworthy is Qonaar Security Systems, Incorporated, of Elk Grove Village, Illinois. This company, too, is basically in another business (parking meters) but is entering the residential security field with a national sales force of what amount to door-to-door salesmen.

Qonaar's basic component is the $375 Receiver Control, which contains an ear-piercing siren. It produces a pulsed tone for an intrusion alarm and a steady tone for a fire alarm. The control has an exit delay after it is armed, allowing the owner enough time to leave the protected area without setting off the alarm. It also has an entrance delay for the same purpose (but it works just as well for a burglar if he knows how to switch it off; he doesn't even have to know in advance where it is because it conveniently lights up automatically when the door is opened). It works on house current but has a standby rechargeable battery in case of power failure.

The wireless intrusion sensors, mounted at doors and windows, cost $40 each. They are also available with individual on-off switches ($45 per switched sensor) so that the householder can deactivate a specific entrance while maintaining security at all other openings. There's even a $40 portable emergency sensor the owner can carry with him anywhere around the house, including out into the yard, or he can leave it at the bedside as a panic button. Heat sensors, which transmit an emergency signal whenever temperatures reach 135° F., can be added to the system at a cost of $40 each. To complete the system, Qonaar can supply a $275 telephone dialer and/or a $50 remote siren to attach to the outside of the house.

The Linear Corporation, Inglewood, California, also makes a good wireless burglar alarm system. The main difference between Qonaar and Linear is that Linear's main business is the manufacture of remote control garage door openers. As in all radio control security systems, the open-circuit system is unsupervised, including the battery-powered transmitters, which should be tested frequently to minimize chances of failure.

Of all the false-alarm-prone radio control burglar alarms, the nearest to being foolproof is the type from Transcience, Stamford, Connecticut. Instead of relying on only two or three signal elements at most to lock out false alarms, the Transcience receiver can be activated only if six separate signal elements are present in correct sequence (the

Alarm bells should be centrally located so they can be heard all over the house, not just placed where they'll be unobtrusive.

Flashguard shoots off regular photographic flash bulbs at sign of intrusion, the light flash activating the alarm system.

frequency, audio tone, and pulse rate are analyzed on one detection circuit, and a second circuit measures amplitude, width, and time-spacing between the pulsed time bursts). Even so, a complete Transcience installation still costs only $400 or $500 on the average, including the self-contained electronic howler.

The double-circuit, six-element receiver is also used in the Preventor burglar alarm from Defensive Instruments, Incorporated, Pittsburgh, Pennsylvania. However, this fast-moving company's ultrasonic system (see below) has overshadowed the radio control system. In fact, the company has spawned a number of spin-offs; when it occurred to somebody that a burglar alarm could be activated with ordinary flashbulbs, for a genuine exclusive in the low-priced security field, the firm of Flashguard, Incorporated, began business almost immediately.

A Flashguard burglar alarm system utilizes regular Sylvania Magicubes, the same as used in cameras. Each intrusion sensor at the doors and windows is armed with two flashcubes (two because any single flashbulb might fail to go off). When a door or window is opened, the sensor's bright flash triggers a monitor that sounds the alarm. The alarm monitor can cover any number of flash sensors in a direct line within fifty feet.

As Don Horowitz, president of Flashguard, likes to tell his rapidly growing army of door-to-door salesmen, the sensors not only require no wires but also require no batteries to fire Sylvania's Magicubes. Any number of flash monitors can be hooked up to a control center, which handles the actual sounding of the alarm. The whole works is in the $300 to $400 price range for the average-sized residence. Just don't let the viewing eye of that monitor get dusty; it doesn't take much to "blind" it.

Photoelectric Eyes

The more mundane photoelectric eye is not much used

Sylvania's Magicubes are easily replaceable in the system made by Flashguard . . . and are usually enough alone to scare off the average intruder.

in residential security systems. It is better used for opening doors in supermarkets than for setting off burglar alarms. One of its drawbacks is that it is too visible. If a burglar sees a ray of light guarding an entrance, he can step over it or duck under it. If he suspects the presence of electric eye sensors, he can smoke a cigarette as he enters, blowing smoke ahead of himself so as to make any light rays more apparent. However, photoelectric eyes can be effective intrusion sensors if used to trigger an alarm at any means of access where its beam cannot be seen.

These electronic controls are manufactured in such volume for industrial use that they are relatively cheap.

IF-22 INVISIBLE FENCE CONFIGURATIONS

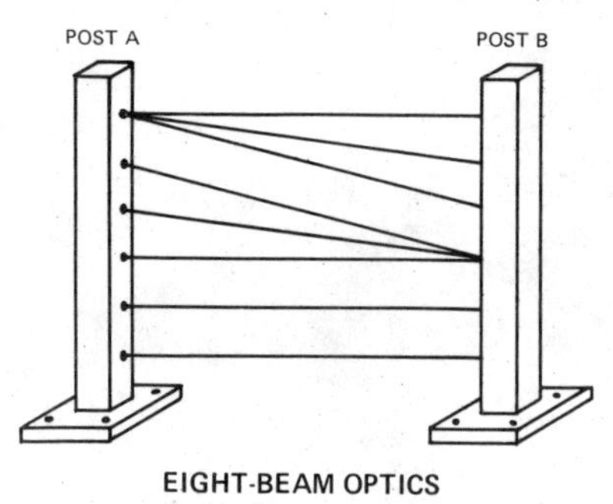

EIGHT-BEAM OPTICS

Two bottom beams may be turned off for heavy snow conditions.

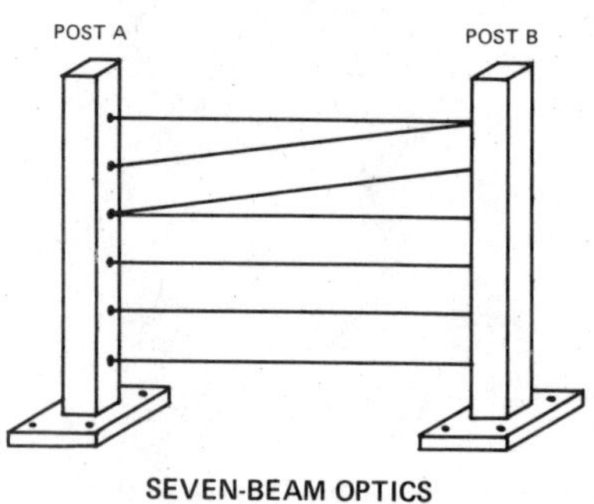

SEVEN-BEAM OPTICS

Three bottom beams may be turned off for extra heavy snow conditions.

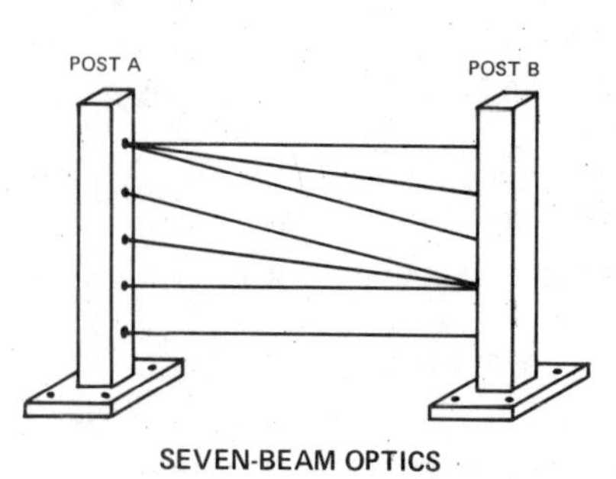

SEVEN-BEAM OPTICS

Bottom beam may be turned off for snow conditions

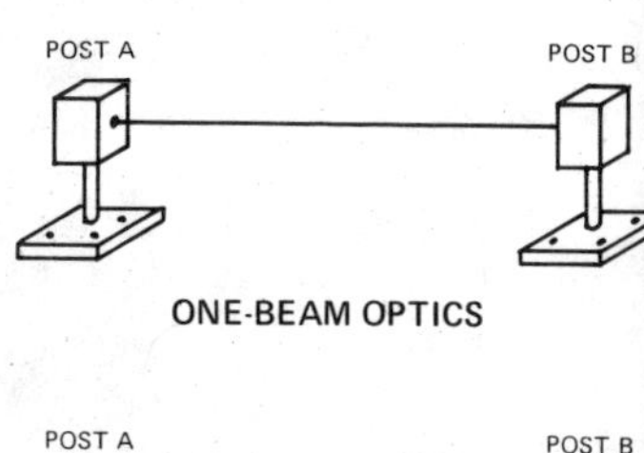

ONE-BEAM OPTICS

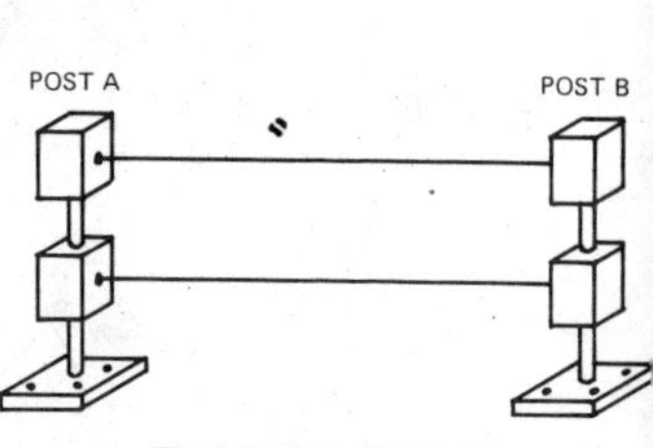

TWO-BEAM OPTICS

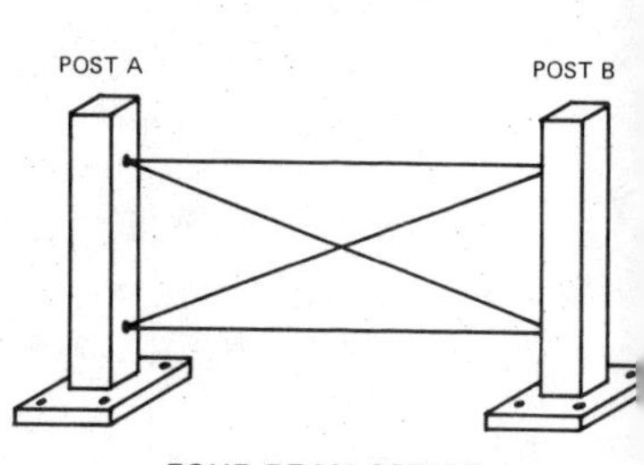

FOUR-BEAM OPTICS

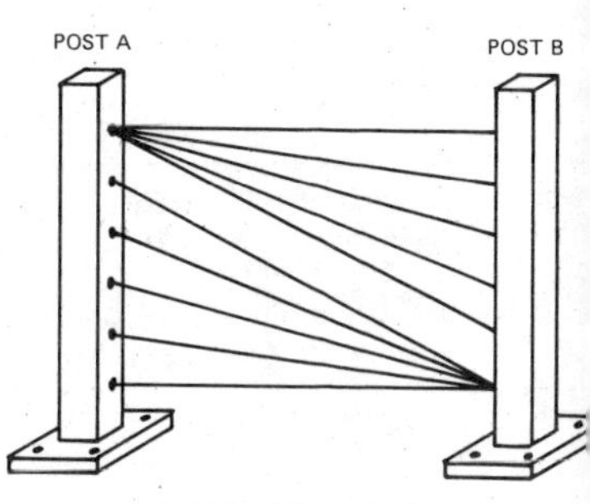

TEN-BEAM OPTICS

Sophisticated photoelectric systems, invisible to the human eye, modulate infrared light pulses to create "electric fences" using up to ten-beam optics.

Autotron, Incorporated, Danville, Illinois, one of the world's leading manufacturers of photoelectric controls, sells complete retro-reflective photo controls with self-contained sensors for as little as $60 in transistorized plug-in models with sensitivity adjustment for varying light conditions. Photoelectric eyes are often used for exterior perimeter alarms, with ranges up to 1,000 feet.

Today's more sophisticated photoelectric systems utilize modulated infrared light pulses that are invisible to the human eye. The O.I.D.C. organization, Santa Clara, California, custom-builds an Invisible Fence, using up to ten-beam optics, which can be virtually impenetrable without triggering an alarm. The self-contained, solid-state, all-weather system has an expected lifetime of ten to thirty years, and the system is so dependable that O.I.-D.C.'s primary business is now servicing government agencies such as the U.S. Atomic Energy Commission and the U.S. Secret Service.

Ademco, among other manufacturers, makes an infrared photoelectric system that looks for all the world like a pair of ordinary wall outlets for electrical appliances. The transmitter and receiver are in one unit disguised as a single-gang wall outlet, and the "bounce back" reflector is disguised as a double-gang (four-plug) wall outlet. They can be recessed in any opposite plaster, wallboard, or paneled walls from five feet to fifty feet apart.

Energy is supplied by a small control pack that can be located anywhere nearby—in a closet, basement, or any out-of-the-way location. The pack is connected to regular house current through a small, low-voltage transformer. Built-in rechargeable batteries take over for eight hours in case of a power failure. The $100 system (cost to a professional installer) is also available in an "AC only" economy version without the standby power feature.

Ultrasonic Motion Detectors

Along with pulsed infrared photoelectrics, postwar

technology also introduced the use of ultrasonics to the security field. An ultrasonic motion detector transmits inaudible sound waves in a pattern averaging twenty-five feet long and fifteen feet wide. These sound waves bounce back from a protected area to the unit's receiver, which analyzes them constantly. Any change in the wave pattern—such as an intruder walking into it—triggers the built-in alarm. There's no way an intruder can get over, under, or around this field of silent sound waves saturating the protected area. An ultrasonic motion detector is extremely simple to install; the buyer simply plugs the small, usually attractively styled cabinet into a wall outlet.

One of the system's limitations is that it can protect only a single zone. Another is that it is prone to false alarms; it can be set off by anything from an electric fan to a curtain blowing at an open window. Some of the tuned circuits used in ultrasonics are so inherently unstable that they can also be triggred by "ambient noise"—frequencies such as from hissing radiators, an air conditioner, or a ringing telephone.

One of the most reliable ultrasonic burglar alarms is from the Advisor Security Division of Aerospace Research, Incorporated, Brighton, Massachusetts. The Advisor III is a $200 wall-mounted 11 x 7 x 2½-inch control sensor that looks like an ordinary stereo speaker. It is a self-contained unit with transmitting and receiving transducers, signal processing circuitry, standby battery with charger, alarm relay output for either local or remote hookup, and tamper-proof connections. Advisor III single-zone security can also be expanded with the use of $65 "satellite" transceivers, matching the Central Control Sensor in appearance, for a cabled-together multi-zone system.

Any or all (up to four) Advisor III monitors have variable pattern adjustments that can direct the sound pattern to the right, left, or downward, or can broaden the normally egg-shaped pattern to almost fifty feet in width, which will, however, cut down the range to as little as twelve feet. An on-off rocker switch at the base of the central control sensor acts as a test light to check for operating performance.

The higher the frequency on an ultrasonic motion detector, the less likely it is to set off false alarms due to ambient noise frequencies. The one made by the Security Devices Division of Systron-Donner, Dublin, California, for example, is much more likely to go off accidentally with its operating frequency of 19.2 kHz than the one from Ademco with a 40 kHz frequency. Ademco's price is only $120, but that's the price to an installer, not to the user.

The Ultrason system from the Alarm Products Division of the Master Lock Company, Milwaukee, Wisconsin, is unique in that it turns on the house lights when it is tripped, and the loud screeching alarm does not go off until thirty seconds later. The sounding alarm itself is a remote satellite unit, but it does not have to be wired to the basic Ultrason motion detector; it is simply plugged into any standard 120-volt outlet, and receives the alarm signal through regular house wiring. Master's satellite alarms can also be used with the company's electronic fire alarms, which are photoelectric smoke detectors that sound an alarm before any heat is built up from spreading flames.

Pittsburgh's Defensive Instruments enthusiastically tells its salesmen that its $400 ultrasonic plug-in system is easy to sell because the prospect can see how a demonstration model actually works before he buys it, not just afterwards as in the case of hard-wire installations. Although only a couple of years old, the company is already considered the biggest operation in the house-to-house security sales field.

Microwaves

Many of the problems inherent in the relatively unstable ultrasonic motion detectors are eliminated in the

Ultrason motion detector has unobtrusive cabinet that fits into most decors, is simply plugged into any standard 120-volt outlet, and does not need to be wired into the satellite sounding alarm.

otherwise comparable microwave plug-ins. Whereas the inaudible sound waves from an ultrasonic device are limited to a single room, the radio microwaves are not confined by normal walls or partitions. A microwave (radar) motion detector will cover up to 3,000 square feet of space surrounding it, but it can be controlled for lesser coverage so that, for example, a dog walking by outside, but through its field of protection, won't set off a false alarm.

More importantly, a microwave system cannot be set off by ambient noise frequencies or even by ordinary radio waves because it operates on such extremely high frequencies. The one from Pinkerton's Electro-Security Di-

Alarm bells should be mounted some distance from the sensors to make disarming them more difficult for intruders. Many burglar alarm systems now incorporate fire alarms, although very few have enough fire and smoke detectors for real protection. Homeowners can also get custom cabinetry for motion detectors if they are fussy about their interior decorating.

vision, Webster, Massachusetts, for example, operates on a 915 MHz band, far out of reach of ordinary waves.

Microwave plug-ins are not cheap—the base price of Pinkerton's Minuteman II is $450. Yet, for that price, the homeowner gets a system in which an initial "lights on" alarm is activated when approximately three steps are taken within the protected area. If motion within that area continues, the second alarm activates a loud, piercing built-in siren after a twenty-second delay. The siren will continue to sound until the intruder leaves, but it will shut off automatically sixty seconds after motion ceases.

The Minuteman II can be turned off only with a special key. It can't be picked up without triggering the siren. It switches automatically to an internal battery in the event of 110-volt power interruption, so that an intruder accomplishes nothing by pulling out the plug. The high-capacity sealed storage battery will operate the unit up to three days in the absence of primary power. A built-in automatic charger keeps the battery at full charge at all times during normal AC operation.

Operating cost is minimal—about 1¢ a day for the small amount of electricity it uses. The Minuteman II can also double the area of protection with the use of a $150 remote antenna. A remote keyswitch (another $35) enables the owner to arm or disarm the unit from a point outside the protected area—at a front or rear door, for example, or from an adjoining room. A remote siren, with tamper-proof cable and plug, costs another $150. The fifteen-pound 13 x 8 x 5-inch console cabinet and external components are all electronically supervised.

Defender Enterprises, Detroit, Michigan, makes a tabletop microwave plug-in which is sensitive only to moving objects weighing sixty pounds or more. No cat can trigger this one, and a good thing, too, because the initial alarm activates a scary strobe light. If the pulsating glare of the flashing strobe isn't shut off within twenty seconds with a key, the siren alarm will start to sound continuously until the intruder leaves, after which the alarm will shut off after one minute. The $500 Defender also includes a portable module the owner can carry in and out of the house, which can activate the siren by remote control at a touch of the panic button. An optional $90 fire, heat, and smoke detector will also activate the Defender's siren.

Microwave motion detectors are considered gimmicky by the hard-wire adherents. Even Bowmar Instrument Corporation, Fort Wayne, Indiana, which uses ultrasophisticated digital combination locks for arming and disarming their system, uses hard-wire door and window switches, along with pressure-sensitive floor mats, as alarm sensors.

$500 Defender model is sensitive only to moving objects weighing sixty pounds or more, to minimize false alarms; it sets off flashing strobe light, and if the light isn't turned off within twenty seconds by the secret switch, the remote siren starts to sound.

Controls

The heart of a $400 Bowmar security system is the push-button keyboard mounted outside the entrance. When the five-number combination is punched in proper sequence on the ten-number panel, the intrusion alarm system is turned off and the householder can enter when the green light goes on. Once inside, pushing a single "on" button at the wall-mounted Convenience Control Panel will arm the system again.

The intrusion system can be disarmed without interrupting the fire warning system, which remains on at all times. The burglar alarm can be shut off from inside at the control panel only by using the same combination of numbers on a duplicate ten-button keyboard as used on the outside entrance keyboard (see Chapter 1 on changing the needed combination number at will).

A supervisory light on the control panel tells the householder when all monitored doors and windows are properly closed, so that he doesn't have to worry about setting off a false alarm when he pushes the "on" button. A separate mat button allows the householder to turn on the floor mat sensors separately—sensors that are normally used when the house is left unattended. An exit button gives the user a sixty-second time delay to leave the house before the intrusion system is automatically reactivated.

Bowmar's security system is powered by standard 110-volt AC house current and uses no more electricity than the operation of an electric clock. A 12-volt standby battery keeps the system operating in case of blackouts or power failures. Bowmar is trying hard to get builders to install its security system during construction, but has a hard row to plow in that cost-conscious field. The company is also attempting to establish distribution of do-it-yourself outfits through lumber yards and building materials dealers.

One of the simplest control panels is the basic one used by Thomas Industries, Incorporated, a major installer of

Bowmar's interior control panel, with push-button combination, monitors all doors and windows for proper closure before the system is armed by anyone leaving the house.

burglar alarms headquartered in Louisville, Kentucky. It consists of just two switches with accompanying lights. When the top one is flicked, a white light at its left goes on if all monitors at doors and windows are in "ready" condition. If so, the bottom "on" switch activates the alarm system and its corresponding light goes on. A comparable three-switch panel also has a red "on" light to show when the fire protection system is in operation.

Hard-wire Thomas installations start as low as $250 for a three-bedroom house. The loud alarm signal, usually located in the kitchen where it is furthest from the front

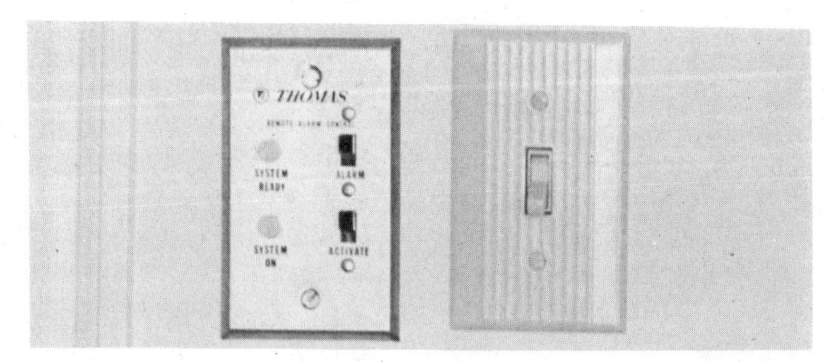

Simple two-switch Thomas control panel has a white light to the left of the top switch, to indicate that all monitors at doors and windows are in "ready" condition. Bottom switch can then be activated to turn on the alarm system.

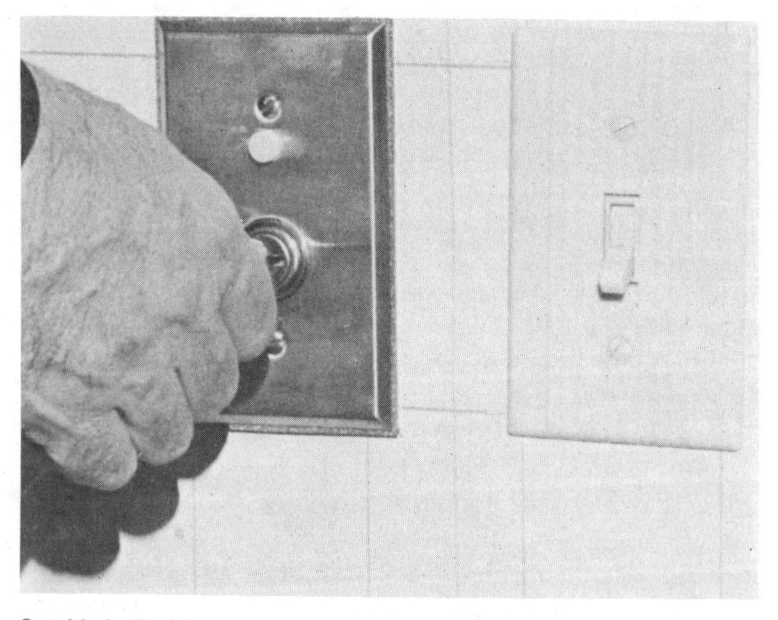

Outside lock and key system allows householder to enter without setting off the burglar alarm built by Thomas. Tamper-proof lock automatically sets off the alarm if anybody tries to take out the screws or otherwise attempts to remove the face plate.

Valuable accessory for Thomas security system is outside tamper-proof combination signal that operates flashing light and loud howler if monitors are activated.

door where forced entry is most usually attempted, sounds when a protected door (or window) is opened. Closing the door or window will not stop the alarm once it starts; the siren has to be turned off at the central panel. Optional remote control switches can also be installed anywhere in the house, as in a master bedroom, so that the alarm can be sounded if anyone hears an unusual sound.

Thomas uses an outside lock and key arrangement to allow the householder to enter or leave the house without setting off the alarm. The tamper-proof lock automatically sounds the alarm if anybody tries to loosen the screws or tries to pry off the face plate. All basic connections are made with lightweight, low-voltage wiring, which makes installation so simplified that almost any competent do-it-yourselfer can wire up the security system himself (depending on the severity of local electrical codes). The only exception is the 110-volt line to the system's transformer.

One of Thomas's best security features is an outdoor light and sound signal combination that flashes and howls

Compact Centralarm gets signals from monitor-connected sensors at doors and windows transmitted over standard house wiring.

when the alarm is triggered. Within sight of the neighbors, it can convince the average burglar that there are easier pickings elsewhere. It has built-in tamper-proof factors, and it will trigger the general alarm itself if anybody monkeys around with it.

Not all hard-wire systems require in-wall wiring (even though it's neater). Master Lock Company's Centralarm control center, the size and shape of a compact radio, sits on top of a table or desk, where its receiver can get signals sent over the house wiring from sensor-connected transmitters. The Centralarm can also be wired up to sensors directly with a minimum of trouble, as each intrusion and fire detector is pre-wired with ten feet of cord and a snap-on connector.

Do-It-Yourself Kits

Master Lock has taken the one-package do-it-yourself

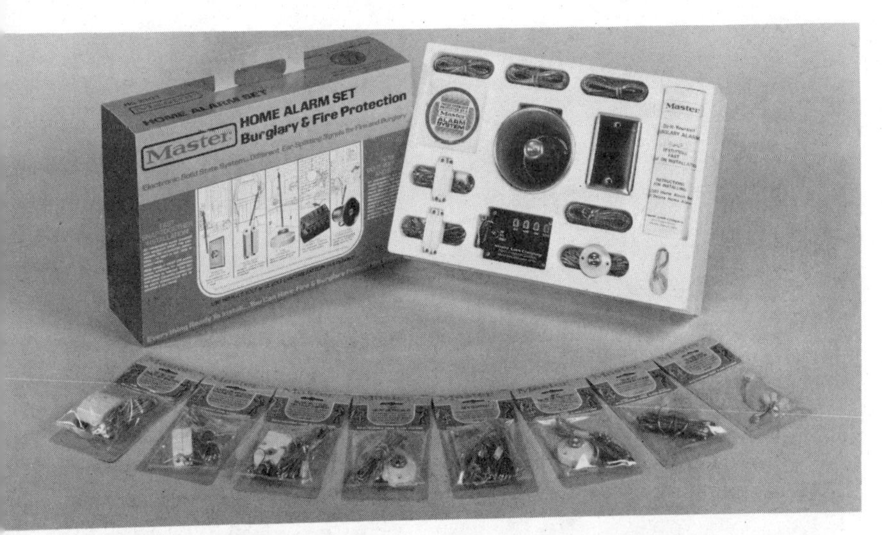

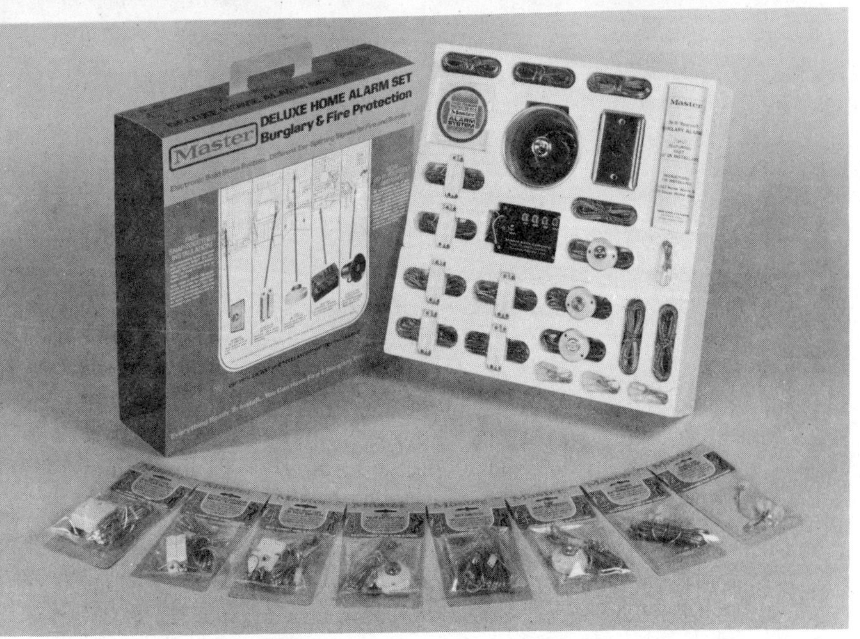

One of the best, though far from ideal, do-it-yourself kits is the package distributed by Master Lock Company through hardware and department stores and discount houses.

kit right down to and including the hardware store market. The kit comes in two standard sizes, and even the $60 kit includes fire as well as burglary protection. However, the kit includes only a single 135° F. heat sensor—of limited value since fires cannot be depended upon to start only in one given location in a house.

The $60 kit does not provide for adequate burglary protection, either, containing only two intrusion sensors. The sensors could be used for the front and back door, or one door and a single window, but only a very small apartment could be adequately covered. However, the little kit does include a palm-sized "control center" and a two-tone siren, along with a key switch so the system can be turned on or off from outside the house. The fire alarm circuit is on at all times, even when the intrusion system is turned off. There's also a three-way connector and four twenty-foot extension cords in the kit, plus an instruction manual and a warning decal. The user has to furnish his own 6- or 12-volt lantern battery.

The deluxe kit, at $90, has six door or window switches, three heat detectors (still not enough for the average-sized house), six twenty-foot extension cords, and four three-way connectors, plus all the rest of the equipment contained in the $60 kit. Additional components are, or course, available as open stock accessories, as are pressure mats, panic buttons, sensors with individual shut-off switches, and remote sirens.

The H. W. Crane Company, Maywood, Illinois, sells a *one*-sensor burglar alarm "system" for $60 (to distributors, which would make it about $90 from a retailer). It consists of a single 30 x 20-inch mat switch, sixty feet of low-voltage wire, and a control box with a self-contained wailer. (Battery not included.) This is definitely not the type of security system that will lower insurance rates for the user.

Stepping on Crane's mat not only trips the siren but also turns on house lights plugged into the control box. A

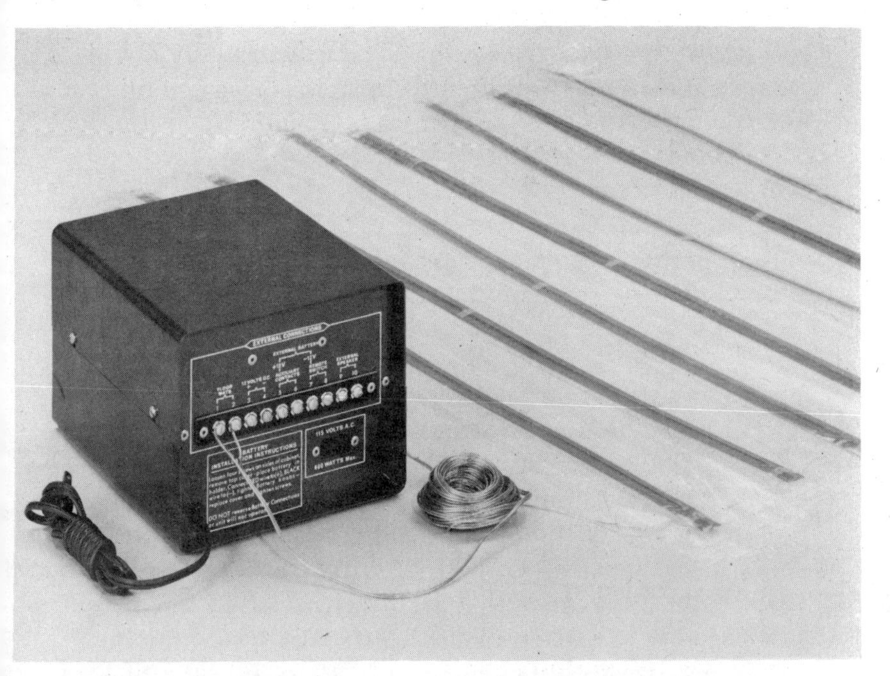

Single-sensor burglar alarms, such as the one made by Crane with a sole mat switch for protection, have limited value for meaningful protection in all but the smallest homes.

neon indicator light on the control box tells when the system is on, for people who don't have sense enough to look at the switch to see whether it is up or down. Extra mats retail for $10 apiece ($5 for stairway mats measuring 30 x 6 inches), magnetic-mechanical switches for doors and windows are $4, as are 135° F. heat detectors, and a remote control switch for turning the system off and on again anywhere in the line is $1.50. A sixty-foot roll of burglar alarm wire is $3.00.

The main trouble with battery-powered alarm systems is owner neglect. If the battery isn't tested periodically, it might be dead by the time an emergency situation does occur. Remembering to check the power every few months is not easy, and the best way to do it is to circle the calender dates when inspections should be made, every time the homeowner gets a new calendar.

Mat Switches

The most inconspicuous mat switch for use under rugs and carpeting where an intruder is likely to step is the Secur-Step, made by the Recora Company, Incorporated, of Saint Charles, Illinois. This wafer-thin mat, less than an eighth of an inch thick, will activate an alarm with any pressure of fifteen pounds or more (it may be safe from the cat, but not the dog). Secur-Step matting is made of durable, moisture-proof vinyl with heat-sealed construction, for low-voltage circuits.

A twenty-five-foot roll of Secur-Step, thirty inches wide, costs the distributor $50 to $60, depending on the quantity he buys. Any running length required for installation can be cut easily with scissors (the original width must be retained). Recora's individual Secur-Step mats are relatively expensive, though; the popular 30 x 18-inch size usually retails for $13.95, and the 24 x 7-inch stairway mat for about $5.

Ademco's line of security equipment includes top quality magnetic-mechanical sensors for doors and windows (as well as mats, foil take-offs, and spring, leaf, plunger, vibration, and mercury contacts), but the company also supplies installers with low-cost but unreliable "reed" contacts that can either set off false alarms or fail to operate altogether. The hermetically sealed gold-alloy-plated reed contacts may be a good choice for damp locations where the ordinarily more reliable all-mechanical magnetic contacts cannot be used. But if installed on infrequently used doors or windows, all reeds are prone to "freezing" due to residual magnetism. If a reed freezes in the closed position, the alarm won't operate in an emergency.

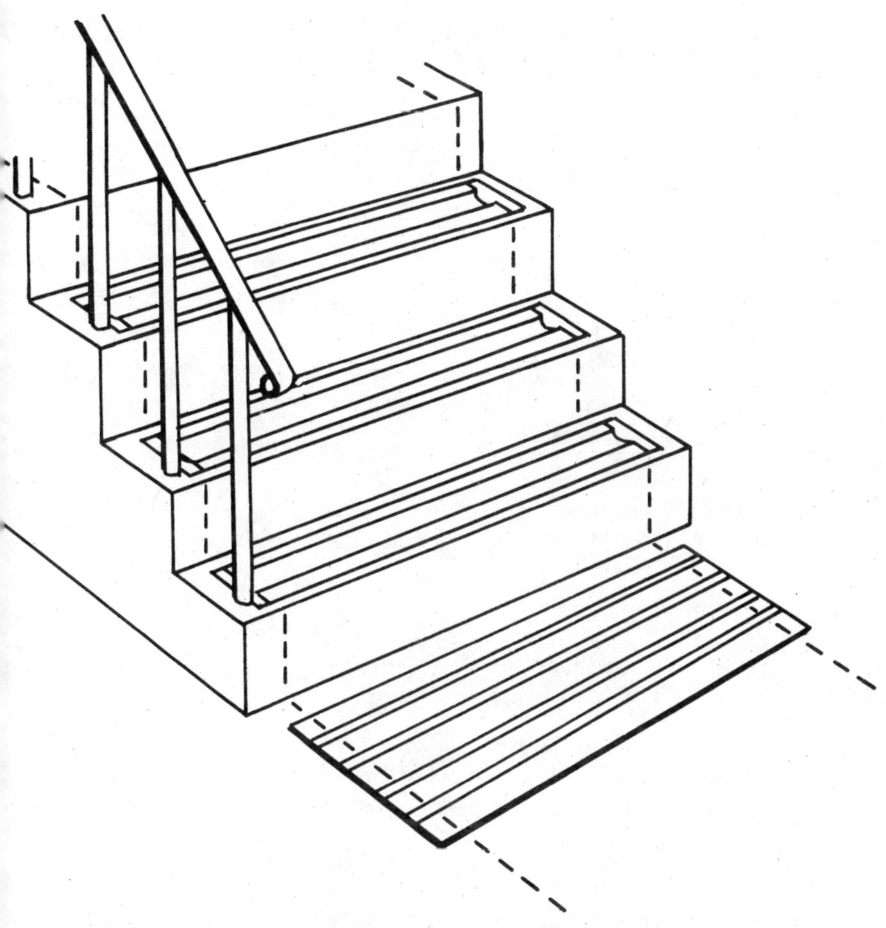

Mat switches set off the burglar alarm if anybody steps on the rug covering them; they set off fewer false alarms than the complex electronic motion detectors.

Inasmuch as all reeds have weak contact pressures, they are susceptible to false alarms caused by vibration. The weak contact also reduces the life expectancy of the magnetic contacts because surge currents, which often occur when the control instrument is turned on, can cause pitting in the contact points.

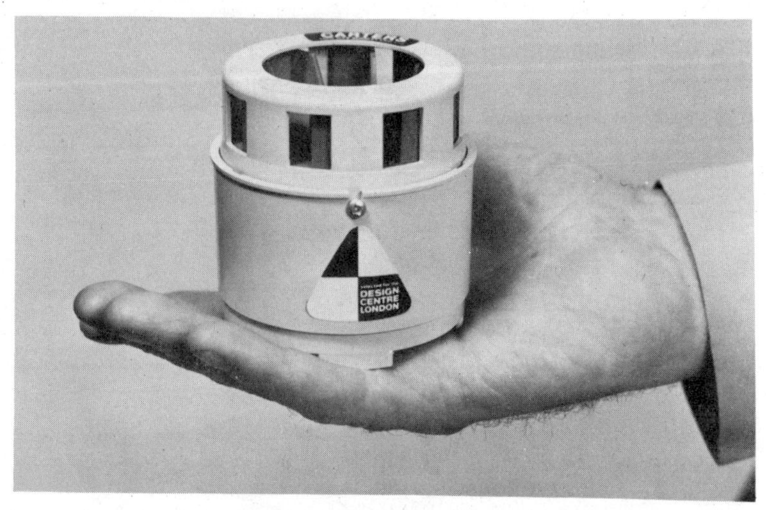

Low-voltage Minimite alarm belies its small size by making twice as much noise as other sirens that are twice as big.

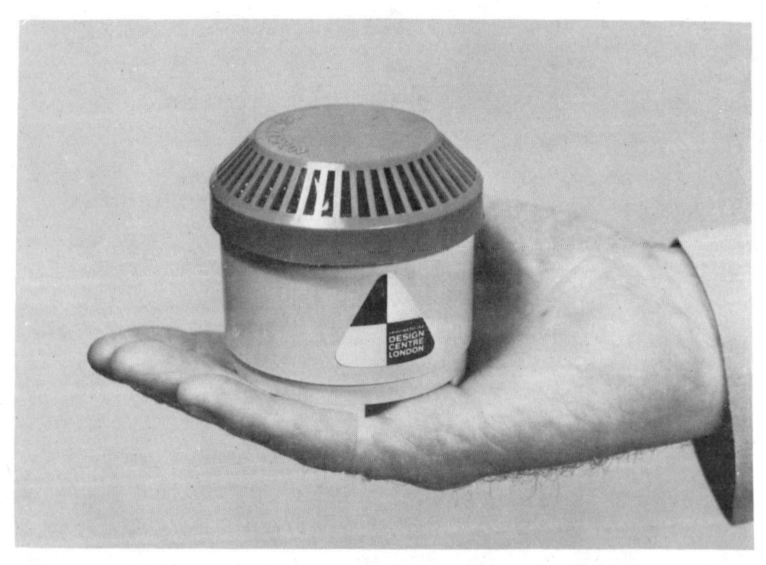

Minitron, with fewer moving parts than the Minimite, is an electronic howler with exceptional durability.

Conventional trumpet-type burglar alarm made by Carters, the largest manufacturer of sirens in Europe, is built of heavy-duty synthetic resin for extra long life.

Audible Hardware

Don't expect to get something for nothing. A good ten-inch outdoor burglar alarm bell alone can cost $35, such as the top grade ones made by the Electronics Division of the W. L. Jenkins Company, Canton, Ohio. That's not counting the housing or the cost of labor for installation. Prices for weatherproof sirens of good quality start at $100.

Among low-cost compact alerts, the best is made by Carters, Europe's largest manufacturer of sirens. The palm-sized low-voltage Minimite measures only 3" x 3", but it makes twice as much noise as many six-inch gongs. The sound level is an ear-shattering 104 decibels at ten feet under outdoor conditions—and even louder indoors. The Minimite is also made in a mains-voltage model, about 2½ inches taller. The rotating parts in a Minimite are all sealed, and the body and shell are made of glass-filled heat-resistant nylon so that it requires no maintenance.

Carters also makes a palm-sized electronic howler, called a Minitron, in both AC and DC models. Because it has fewer moving parts, the Minitron is even more durable than the Minimite. In addition, the conventional trumpet-type yelper from the Lancashire, England, company is made of extra-heavy-duty synthetic resin and is acknowledged as the longest-lasting in the industry.

All good security systems should have UL-approved components. But in hiring a professional installer to put in a good burglar alarm system, the most important single thing a homeowner has to consider is the installer's reputation. He should be checked out through the Chamber of Commerce, the Better Business Bureau, and Dun & Bradstreet (if he is financially insecure he will be tempted to make a killing on each job he gets). He should also be checked out with satisfied customers. If the company doesn't stand up under that kind of scrutiny, its burglar alarm products are no bargain at any price.

7

Community Relations

"The only time I ever heard of a bunch of Republicans acting like Communists is the way those friggin' people up in Evanston watch each other's property like they all owned everything together. Nobody just minds his own business"—Inmate, Cook County Jail, Chicago, Illinois.

Unless the best burglar alarm system in the world is hooked up to a central station or to the police station, its effect is minimal when the family is away if the neighbors do not recognize the alarm and take action accordingly. Any time a burglar alarm sounds, whether the family is home or not, neighbors should call the police immediately. Nobody should ever assume that somebody else is calling the police; they do not mind getting duplicate calls, and they universally say that answering duplicate reports is far better than getting none at all.

One of the best possible methods of residential security, with or without burglar alarms, is to get acquainted with the neighbors. You don't need to be drinking buddies or go on family picnics together; just get to know them.

Go across the street or next door and say: "Burglaries are getting pretty bad around here, so if I see anybody suspicious fooling around at your doors or windows I'll telephone to tell you, and if I don't get an answer I'll phone the police. Will you do the same for me?" Who would say no?

The fact that very few people in high-rise apartment buildings know each other is one reason for the inordinately high crime rates in such buildings; nobody pays any attention to what anybody else might be doing. The dozens of deliverymen and repairmen constantly going in and out also help make illicit entry easy. Neighbors keeping an eye out for each other would go a long way toward making such buildings a lot safer.

The police are glad to respond to any "suspicion" calls, even if they turn out to be false alarms. In fact, most police departments would rather have alert citizens phone them directly when their suspicions are aroused, instead of trying to reach an unsuspecting householder first. Except in the biggest cities, a patrol car can usually be depended upon to appear at the scene of a suspected "crime in progress" within three or four minutes—and the police have the training to handle the situation without shooting anybody.

The number of ostrich-like people who "do not care to get involved" is smaller than some newspapers would lead their readers to believe. Most people *do* care about their fellow man. Newcomers to big cities are often shy and do not approach their neighbors because of fear of rejection. Usually, all it takes is somebody making the first move. It might as well be you.

Neighbors do not need to be busybodies, and there is no point in trying to be a hero when police are only minutes away. People who see or hear anything suspicious in the neighborhood should not investigate personally but should call the police.

Householders should know their neighbors' cars as well as members of their families. The presence of strangers in the area should always alert neighbors to potential trouble; the strangers may or may not be legitimate visitors. Jotting down the license numbers of strange automobiles can do no harm even if the drivers are innocent as lambs, but the numbers can turn out to be helpful to the police in many cases.

A ten-year-old boy dawdling over his homework one night in the Webster Groves section of Saint Louis noticed a strange car parked outside his house with the motor running. Being home alone, he called in the license number to the cops, who put it into the computer and got a report within seconds that the car was indeed a stolen vehicle. A patrol car arrived at the scene just as the suspect car was pulling around the corner, to pick up a burglar-accomplice who had been looting a neighbor's house, with the two thieves having maintained contact by walkie-talkies for a getaway that had been planned for split-second timing.

It's always a good idea to let neighbors know when you leave and when you expect to get back. If you leave for a wedding ceremony that will take a couple of hours (often announced beforehand in the newspapers for the convenience of burglars), and the neighbors see lights going on and off in the rooms half an hour after you have left, they will know that something's wrong and can call the police.

911

At one time, community-minded citizens had to have not only the phone numbers of all their neighbors but of the local precinct station as well. Today the growing use of the nationwide 911 phone number, starting in New York City, goes a long way toward simplifying emergency calls to the police. In other big cities such as Chicago, one central phone number gets faster action through the emergency clearing station than through local stations; the officers on duty at the central switchboard have citywide information at their fingertips covering the locations and types of all emergency facilities at all times.

MO Recognition

A Miami housewife noticed a stranger putting glass panes *back into* a louvered door in the breezeway of the house next door. These jalousie-type windows are notoriously hard to burglar-proof because the individual glass panes can be lifted out so easily (unless they are glued into the slides with an application of two-part epoxy resin). Her call to the police resulted in an immediate contact with a police car only a couple of blocks away, and the burglar was nabbed before he could even leave the yard. The police even knew who he was before they got there, thanks to an en route computer clearance. This burglar was known to the police for his M.O. (Method of Operation)—specializing in entry through louvered windows, doing a tidy C&D job (strictly for cash and drugs), and then covering his tracks so that returning householders might not realize that their house had been burglarized until days afterwards, if ever.

Most burglars who get away with one certain method of operation two or three times continue to use the identical method until they are finally caught, setting up an identifiable pattern that police can recognize as the work of one individual. Many convicted burglars even continue the same M.O. *after* they get out of prison to resume their trade.

One M.O. that results in a lot of people not realizing that they have been burglarized is the Monday Scam. Originally, thieves specializing in this type of crime would watch for housewives hanging up their laundry in the back yard on the traditional wash day. Many people do not bother to lock up the rest of the house as long as they "are home," although busy in the backyard and basement. The burglar simply walks in (or maybe cards the front door latch), neatly cleans out what he wants in a few minutes, and unobtrusively leaves with everything apparently undisturbed. When the housewife does go to look for her wristwatch or her house money, anywhere from hours to days later, she most often figures that she has misplaced it (or blames somebody else in the family). The so-called

Monday Scam is used every day in the week, for example, when the resident is in the yard doing gardening, supervising the children at the backyard pool, or gabbing over the fence with a neighbor.

A related M.O. is watching for parties when guests all use one bedroom to stash their coats and purses; when everybody is having a good time at the backyard picnic table or in the recreation room downstairs, a sneak thief can rifle the purses in a matter of moments. Who counts their money before they go home from a private party?

This type of M.O. gives the burglar a great deal of time to get away and to dispose of his loot. When the people do finally realize that their property has been stolen, they often fail to report the theft to the police at all, for fear of seeming somewhat foolish. Some police chiefs at their 1974 convention stated they believe that less than a third of residential thefts are actually reported, which makes their job of crime prevention considerably more difficult.

In the interests of community welfare, anybody who has been burglarized, regardless of the amount of the loss, should report the incident to the police. Every bit of information helps to establish recognizable M.O. patterns. Some burglars take only certain types of property (brand name wristwatches, but no custom-made jewelry). Some specialize in rifling medicine cabinets for amphetamines (speed) or barbiturates (goof balls), which are legitimately used as prescription drugs by millions of solid (if not stolid) citizens. One individualist had thousands of Nembutal capsules in his room when caught, explaining that he didn't take anything else because it was too dangerous.

One Monday Scam operator was identified as a careful thief who invariably unhooked the bedroom window screen so he could have a fast getaway in case anybody approached while he was at work. Another always got in by slitting a screen above the catch and then using a piece of Scotch Tape to patch the slit, so it would be unnoticed for as long as possible. Most burglars develop idiosyncrasies like these, and a whole raft of burglaries can often be cleared up when one of them is solved.

The citizen who has been victimized can help a lot by reporting any and all thefts to the police, with complete details. The information not only helps the police catch burglars but also makes it harder for even a good lawyer to get a burglar off as "a first offender who made a mistake on impulse," when it turns out that the miscreant has pulled a couple of dozen other jobs just like it.

Holiday Hits

Many burglars specialize in "vacation hits" exclusively. To them, the word "vacation" just means that the house will be vacant. A family leaving for a vacation should not only stop newspaper and milk deliveries, as noted earlier in this book, and take sensible precautions to make the house look lived in by having the yard taken care of and using a light timer while away, but should also leave a key with a trusted neighbor. Anybody keeping an eye on the house for a vacationing neighbor should pick up throwaway newspapers and circulars, as well as take in the mail, and change the positions of drapes and shades every day or two.

Some homeowners even hire "house sitters" to live in the house while they are away on vacation, often college students who welcome the chance to live beyond their customary means for a couple of weeks for very little pay (check references before hiring house sitters, though). Naturally, neighbors should be so advised lest they have the students arrested.

Other vacationers have an extension of their burglar alarm installed in the house of a cooperative neighbor during the length of the absence. Installing such a remote alarm costs very little when the electric power supply for both houses operates on the same general circuit.

If vacationers notify the police as to their vacation

plans, the house will get special attention. The homeowner cannot expect the police to come around to rattle doorknobs and check for windows left open. But just their more-frequent-than-usual presence in the neighborhood (even on the spot-check basis most usually used) will discourage burglary attempts. Nobody should expect even the best of neighbors to be on twenty-four-hour guard duty, either, and vacationers should notify the police, too, as to where they can be reached in case of an emergency.

Neighborhood Watch Organizations

Within the last couple of years, security-conscious citizens in a number of cities have entered organized programs in a self-help cooperative battle against the growing number of burglaries. This total citizen participation and involvement is best exemplified in Los Angeles, where the police department under Chief of Police E.M. Davis sponsors the Citizens' Neighborhood Watch.

Department personnel make public appearances throughout Los Angeles to alert civic groups, service clubs, and PTA meetings to what's going on, to give people basic instruction on security, and to tell them when and how to call for fast action at signs of suspicion. The department uses specially marked Community Relations cars, prints dozens of pamphlets and brochures in both English and Spanish, and has a separate Public Affairs Division under the command of Captain D. P. Sheery. Citizens who participate in the Neighborhood Watch program get brightly colored window and bumper stickers to let potential burglars know that the word *Citizen* is spelled with a capital *C* in that particular neighborhood.

Although the Los Angeles Police Department averages about 3,000 emergency calls a day, fewer than 200 are to report burglaries. That's still an awful lot of burglaries, but a burglar thinking about making a heist in an area with Citizens' Neighborhood Watch stickers in all the windows knows that he is taking more of a chance of getting caught than in other neighborhoods.

The dictum in all cities using similar programs is that crime prevention is a responsibility of all citizens, and not just of the police department. Little by little, the country does seem to be growing up in facing its responsibilities and obligations.

In Chicago, the police department has helped organize Block Clubs that meet periodically to compare notes on mutual-security measures. The Chicago police department also developed a Citizens' Patrol, operated by numerous community organizations. Members of the Citizens' Patrol take turns in operating their own private automobiles for street duty, traveling in pairs in high crime areas, often maintaining contact with each other with walkie-talkie radio systems.

The Citizens' Patrol has often been called the eyes and ears of the Chicago Police Department. Since they do not carry firearms, members of the patrol take no direct action themselves but call the police when they see any trouble brewing. They have helped to make Chicago one of the *safer* major cities in America.

If not tightly controlled by responsible supervision, a patrol program operated by unofficial volunteers can degenerate into a vigilante operation in which the cure can be worse than the malady. In actual fact, no member of a citizens' patrol has any more authority than anybody else. He can make a "citizen's arrest," but so can any other citizen. Theoretically, any citizen can arrest the perpetrator of a crime and march him off to the police station, but in actual practice it means holding him until the police arrive. Even then, an untrained citizen risks charges of false arrest. The far more sensible policy is to call the police and let trained officers do any necessary arresting.

People who call the police when they see something untoward in the neighborhood can indeed "get involved."

Getting called down to police headquarters and possibly picking somebody out of a lineup (which is basically for the protection of the accused criminal) can be a bit annoying—and being a witness in court can be inconvenient. But jury duty, or even going to the polls to vote, for that matter, could be viewed as getting involved. It's all part of paying your dues for living in a civilized society. Make no mistake about it, the "War against Crime" is a real war—one that needs every citizen's involvement.

8

Retaliatory Devices

"Okay, I break the law when I jimmy somebody's door, sure, but that doesn't mean that a lousy citizen has the right to do anything illegal to me"—Inmate, County Jail, Los Angeles, California.

Setting a trap for a burglar is a temptation too great to resist for homeowners who have been burglarized repeatedly. But the philosophy that "all's fair in love and war" does not apply to the war against crime. The homeowner who sets a trap to retaliate against an intruder can find that the trouble he gets into himself is not worth whatever personal satisfaction he may get out of it.

Legal Risks

In Wisconsin, a homeowner who was periodically away from his house for several days at a time had been burglarized three times within a few months. He lived alone and was not the best housekeeper in the world (burglars will ransack a disorderly house from top to bottom looking for money, on the premise that less-than-careful people are also careless about putting money in the bank where it belongs). The Wisconsinite's house had thus been vandalized all three times, with dresser drawers emptied, mattresses slit open, and everything in the pantry thrown onto the floor.

The police had been unable to do anything, but keeping a dog locked up in an unattended house for days at a time was out of the question, and the homeowner felt that the house was too remote for an ordinary burglar alarm to be much good. All but gnashing his teeth over the latest outrage, the irate householder rigged up a 12-gauge shotgun pointed at the door, with the trigger tied to a string across the threshold.

He made several trips out of town without incident. Then sure enough: BLAM! He got a phone call at his Milwaukee hotel advising him that another burglary attempt had been made on his house, and that he had better get home pronto.

The burglar was in the hospital with a badly shattered leg from the shotgun blast. Police had found him screaming in a pool of blood with burglar tools scattered all around him. He had a long police record, including several burglary convictions; the car he had been driving proved to be hot; and stolen merchandise was found in his apartment. He confessed not only to the attempt where he had been gunned down but also to one of the other burglaries at the same premises.

There was no doubt about his guilt; he was tried and convicted and was sentenced to prison. By that time, the leg had had to be amputated. Some of the "yardbird lawyers" at the Wisconsin State Penitentiary began advising him of his rights. After he had served his time, he got hold of a Milwaukee lawyer and *sued the homeowner* for damages.

At this writing, the case is still in the courts, but there's a very good chance that the burglar will be able to collect. The homeowner is well off financially and has property worth going after. He has spent considerable time and money defending himself, and will spend more. Incredibly, the case is still being pressed despite the fact that the now one-legged burglar has again been arrested on another breaking and entering charge.

If there had been a sign on the front door reading, "Warning: There's a loaded shotgun pointed squarc at anybody who comes through this door," it might be supposed that the homeowner's position would have been validated. Not so, according to the lawyers, because of the possibility that a prowling burglar might be illiterate.

It is patently illegal to use deadly force against *anybody* except to protect your own life. According to the law, merely protecting property does not warrant the use of any such force. You can legally cause bodily harm to a burglar only if he is carrying a weapon, and even then only if he threatens you with it. The fact that a 6'4" burglar who weighs a well-muscled 240 pounds *is* a threat doesn't count. And if nobody's home, the homeowner cannot legally so much as set a mousetrap that might cause an intruder injury.

A Chicago citizen is currently in even more trouble because he himself was the armed trap. His garage had been burglarized several times, and his motorcycle had been stolen in the latest break-in. He bought a new motorcycle, which he let everybody in the neighborhood know about, staked himself out in the garage rafters the next night with a six-pack and a gun, and left the garage door unlocked.

He was awakened from a snooze at two in the morning by the sounds of the door being stealthily opened, and he peered down to see three prowlers starting to wheel out the new bike. He yelled, "Freeze!" but all three started to run—and he opened fire. One got away clear, but he shot two of them, one of whom died right there in the alley.

After the police arrived, the householder was dumbfounded to find that *he* was in custody, charged with manslaughter. The worst mark against him—entrapment. Although the trio turned out to be the same ones who had been burglarizing his garage (among others) before, the lawyers claimed that he had *lured* them into committing the felony by deliberately leaving the door unlocked.

When a plainclothesman on Detroit's notorious vice squad approaches a streetwalker in the guise of a John and then arrests her in a hotel room for prostitution, *that's* entrapment. But it is sometimes difficult in the layman's mind to equate the entrapment law with a beleaguered farmer who sets a bear trap in his hen house out of desperation and who is thus responsible for breaking a chicken thief's ankle. Legally, though, he is liable. What a property owner should do, say the lawyers, is call the police when a crime is being committed.

There are, however, plenty of ways to set a booby-trap for burglars without shooting them, but which will still make them wish they had gone elsewhere. The classic one is the "practical joke" used by college boys—balancing a bucket of water on top of a door left ajar. The homeowner can fill his bucket with molasses, glue, paint, or alkaline dye, which the intruder will find impossible to scrub off (it has to wear off). This may be all right for garages, sheds, or barns, but can make an unacceptable mess in a kitchen or a living room—and an absent homeowner does not want an infuriated intruder wantonly wrecking the place out of revenge, either.

Identification dyes, and the automatic guns that can spray them, are available from many firms that deal in bank supplies. The nitrobenzene base used in these dyes also stings like crazy, although there are no permanent ill effects. If a dye-marked criminal is caught within seventy-two hours, the identification is absolutely positive because this type of dye is not used in any other applications besides security systems.

Tear Gas

Tear gas is also widely used as a crime deterrent. This

irritant not only causes an excessive flow of tears but is also quite painful to the eyes. The effect is almost immediate upon contact. Regulation tear gas is not sprayed but is released upon the explosion of a projectile fired from a gun with a gunpowder cap. The gas itself is invisible; any smoke you see is from the projectile's explosion (or in gangster movies, from a special effects man who pumps it onto the set). The dispersion of tear gas is very slow, because its weight is very close to the weight of the air itself. In a closed room, it won't disperse for days.

The effects of tear gas are not long-lasting, and any incapacitation usually clears up after an exposure of ten or fifteen minutes in the fresh air. It's a bad ten minutes, though. Anybody who has taken the full force of a tear gas blast is usually ready to call it a night.

There are several formulations for tear gas, but the formula developed by the U.S. Army under the code name CN is the one most widely used, as in riot control. The Army also has a new tear gas, code name CS, which produces instantaneous and almost total incapacitation upon inhalation, whether the eyes are contacted or not. This wicked stuff causes a severe burning sensation of the upper respiratory passages, the nose, and the mouth. It also causes an involuntary inability to open the eyes, plus psychological disorientation, including an uncontrollable compulsion to run.

A third type of tear gas, called a sternutator, also has more distressing effects. Along with excessive tears and burning pain, it produces an immediate headache, sneezing, coughing, nausea, vomiting, and the involuntary fouling of one's underwear. This one can have longer lasting effects for some people, who may require the attention of a doctor and remedial medication.

Tear gas shells are readily available for 12-gauge shotguns, .38 pistols, and the pocket-sized "fountain pen" projectors often sold by mail order companies. The Hercules Gas Munitions Corporation, Chicago, Illinois, makes a tear gas projector that can fire its screw-in .45 shell twenty feet. The pocket-clip firing devices are relatively cheap, and even the highest priced mail order models seldom cost more than five or six dollars.

The shells themselves are expensive, though, from seventy-five cents apiece for .38 shells to three dollars apiece for shotgun shells. They all deteriorate with age, with a life expectancy for dependability of about three years. Many manufacturers such as Smith & Wesson date their tear gas shells for the same reason Kodak film is dated, and they cannot be considered reliable after the expiration date.

Outdated shells should be fired, not just thrown out with the garbage, lest they be discharged accidentally by somebody who might find the live shell. Whenever a tear gas shell has been fired from a gun, for any reason, the weapon should be cleaned as soon as possible because the residue is so corrosive to metal.

If you are considering the use of tear gas, there is one important word of caution. The civilian use of tear gas shells is illegal in Illinois, and mere possession is against the law in New York and California. Many municipalities across the country also have local ordinances making tear gas illegal for ordinary citizens. In some places, special licenses are required. Setting a burglar trap armed with tear gas shells in such areas will do more lasting damage to the homeowner, through being charged with breaking the law himself, than to the gassed burglar, who will usually be completely recovered within fifteen minutes (unless he blindly runs over a cliff after being doused with CS). Anybody considering the use of tear gas should check with his local police department as to legality before buying any.

Mace

The anti-personnel chemical known as Mace is CN tear gas that is ejected under pressure as a spray from the firing device, rather than from the explosion of a fired projectile.

Mace was originally developed by Smith & Wesson for riot control, but the word is now used as a generic term for spray versions of tear gas. Legally or not, a good many women now carry aerosol cans of Mace for personal protection. Many brands combine a dye with the spray, to mark an attacker's hands and face indelibly. One of them, the Guardian "pen" made by Jet Manufacturing Company, Great Neck, New York, is a four-inch plastic tube three-quarters of an inch thick holding a half ounce of a rather mild oleo-resin tear gas combined with a red dye. The chief drawbacks of this device are that it is undated and that its squirt has a range of only three feet, but then, it only costs two dollars.

Considering the relative merits of Mace and tear gas shells, one is forced to admit to the superiority of tear gas shells. A spray can of Mace is much more difficult to rig as a burglar trap than a tear gas shell. A gun only has to be triggered, while the release on a spray can has to be *held* down to get the full charge out. A blast from a pressurized can is not as effective, since it sprays only in one direction, as the explosion of a tear gas shell, which will permeate an entire room with the choking fumes and drive out even the most determined intruder.

Electric Shock

Electricity, circuited to produce an electric shock, is the most common retaliatory weapon in ordinary use. It is mostly used outdoors to electrify fencing, but it can also be extremely effective in wiring up doorknobs, light switches, and all manner of electrical appliances that can deliver a controlled but agonizing electric shock to anybody touching them. Electric shock systems can be operated on either AC or DC current and are most often wired up in conjunction with a conventional burglar alarm system (see Chapter 6). Do-it-yourselfers should *never* attempt installing electric shock systems on their own unless they are quite expert in

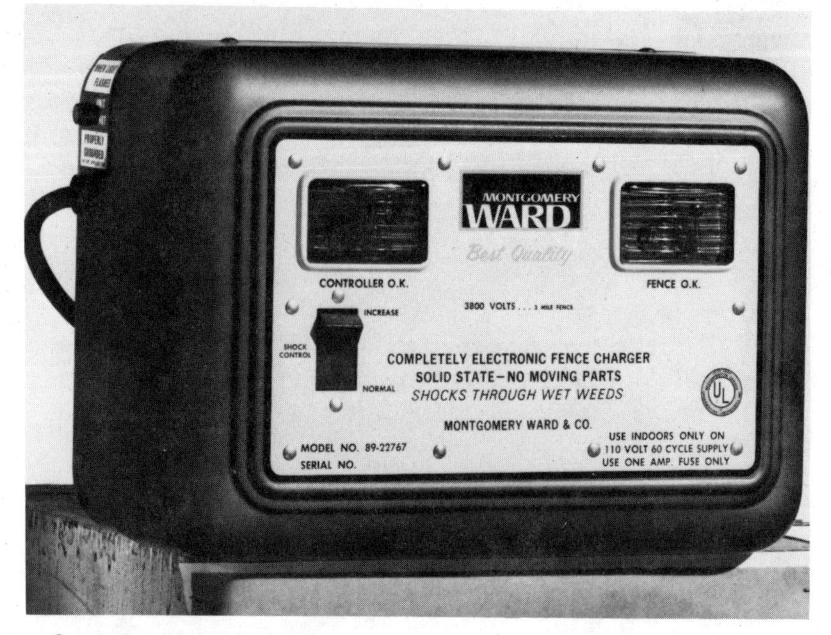

Good electric fence chargers are built of solid-state components for foolproof operation, with no moving parts.

the handling of high-voltage electricity.

Electric shock can be controlled to produce anything from a tingle to a jolt capable of knocking a man unconscious (or even to kill him, if the homeowner is rash enough to use that much power). An electric shock is painful only while the contact is maintained, and there are no adverse aftereffects.

Hazards

Some retaliatory devices can cause so much anguish as to be considered inhumane. A homeowner in Danville, Illinois, who had been burglarized several times, represents a prime example. He rigged up a shotgun trap loaded with tear gas shells but was arrested by the police when they learned of his illegal lash-up, and it cost him a $25 fine. He then poised a string-trip net over the inside of his door, with the net studded with triple-barbed fish-hooks. A subsequent intrusion attempt resulted in the burglar enmeshing himself in the torturous net so painfully with every movement that he himself phoned the police after agonizingly making his way to the phone, begging them to come with wire-cutters to cut him loose.

But the legal risks involved are not the only hazards. Burglar alarms are bad enough when they are set off accidentally, but a retaliatory device going off accidentally can be much, much worse. Any homeowner rigging up a burglar trap of any kind must make absolutely certain that it won't be activated by an innocent person, including himself.

An absent-minded person, or one who is in the habit of drinking too much before he comes home, should never even consider using one. A booby-trap should never be used if anybody has a key to the house other than the person who sets the trap, either; giving the wife a faceful of tear gas and dyeing her bright red does not do much for domestic tranquility.

A neighbor can also get pretty indignant if his dog wanders over and electrocutes himself by lifting his hind leg against an electric fence. So can a meter reader who unwittingly steps into a muskrat trap the householder forgot to disarm.

Putting up signs reading "Beware: This entire property is booby-trapped when unoccupied," is always a good precaution, even though it may not stand up in court.

9

Security Lighting

"I wear dark clothes when I'm working so I'm hard to see at night, but the dark clothes work in reverse for visibility when I'm in a light. I admit it—I am as shy of light as a cockroach"—Convict, State Penitentiary, Lucasville, Ohio.

Most burglars prefer to work in the dark so they cannot be seen. With the mental makeup and social attitudes of the average burglar, this aversion to light is psychological as well as practical. This is one reason that sneaks (including obscene telephone callers) break down so readily when detectives get them into the squad room for interrogation. That bright overhead light spotlighting the suspect in his chair, with his interrogators all but invisible around him in relative darkness, is starkly terrifying for such a criminal.

Most communities realize the value of light as a crime deterrent, and cities such as Chicago maintain alley lighting as well as street lighting. A phone call informing the police that a public lighting unit is out gets immediate attention almost anywhere; a street light may be inoperative because some kid broke it with a rock for sheer devilment, but the police almost always assume that it has been darkened deliberately to facilitate the committing of a crime.

Suburban areas, on the other hand, very seldom have

lighting in the alleys. Most individual houses in residential areas get adequate illumination from the street lights, but the back of the house is usually dark, and often the sides as well. Security lighting in such areas is one of the best investments a homeowner can make.

The homeowner who uses exterior lights as his basic security system is even safer than the person who relies on the noise of a burglar alarm. The burglar alarm chases burglars away *after* they have made an attempt at intrusion, but light inhibits them from even trying. Exterior lighting even serves as a kind of burglar alarm in reverse: when a security light is noticed as being unlit, it alerts neighbors to an unusual situation, and they can phone the homeowner to find out if there's anything wrong. If there's no answer, they can then phone the police.

In the interests of conserving energy, some people who have exterior lighting systems turn them on only when they hear strange noises outside. The prowler who is suddenly revealed in bright light will almost always scuttle away as fast as he can. This sudden action, too, should alert neighbors.

Yard Lanterns

The most commonly used yard lights are designed to be decorative rather than functional. Many have low-wattage flame-shaped or candelabra bulbs that do not throw much light, but they often throw enough. Any yard light is better than none, including Mediterranean-styled lanterns, ersatz coach lights, and Japanese lanterns mounted on posts. The main objective is to have a light—any kind of a light—illuminating areas not covered by street lights.

The post for such lanterns should be set into the ground well away from the house, to provide as wide an area of illumination as possible. A hollow metal post three inches in diameter will accommodate most standard lanterns. It should be rustproof aluminum or coated steel for durability, and the best ones (which cost between $10 and $12) telescope for four- to eight-foot heights. A solid, non-telescoping seven-foot post, prewired for ease of installation, commonly costs $8 or $9.

The lantern itself can be anything from an $8 8 x 15-inch traditional model to a $35 three-bulb showpiece measuring up to two feet in height. A modern white plastic sphere, fifteen inches in diameter and with a 150-watt bulb, costs about $25. People who go in for handwrought

Good yard lanterns should have pre-wired posts for easy installation and should include a weather-proof (capped) electric outlet for use of outdoor power tools and appliances.

ironwork or scrolls can spend almost any amount of money they like, but there are plenty of handsome lanterns on the market, including prewired eagle-crested models with frosted chimneys, for around $12. Solid brass lanterns, for about $20, are more durable than the lower priced steel versions, or even the ones made of cast aluminum.

When buying any lantern, the homeowner should carefully examine the construction. Whether the panels are ordinary glass or break-resistant Plexiglas, they should be removable for cleaning and for replacing bulbs.

Porch Lights

Many yard lanterns are sold in sets with matching wall or ceiling lanterns for the porch. A porch lantern usually takes a higher wattage bulb than the yard version, for the convenience of people using the latch key and of anybody answering the door from the inside who will want to identify callers. Letting a porch lantern burn all night with a 60-watt bulb costs about fifty cents a month for the electricity, which is cheap insurance for the protection it gives the householder (a 100-watt bulb, using almost twice as much power, costs less than a third of a cent per hour to burn).

Wall-mounted porch lanterns, usually extending five to ten inches from the wall for better illumination, cost even less than post-mounted yard lanterns, with prewired coach lanterns, for example, selling for as little as $10 or $15 a pair. Even fancy porch lanterns, such as hooded Early American models with brass reflectors, cost less than $20.

The homeowner who is concerned only with functional illumination, rather than with decoration, can get prewired porch lights for only $2 or $3 if he shops around. Flush-mounted ceiling lanterns are relatively cheap, too. In some ways, they are better than the more expensive chain-hung ceiling lanterns, which bring the light down for better

Wall-mounted porch lanterns should extend well away from the wall to provide maximum illumination.

illumination but which can be broken from swinging in high winds.

Floodlights

Utilitarian floodlights bathe a yard in brilliant light from the big, 150-watt mushroom-shaped bulbs, which are weatherproof and require no enclosures. Their wall-mounted, cast aluminum holders cost about $4 for a swiveling single-bulb unit. Double-socketed holders, for

Low-cost floodlights utilize 150-watt mushroom shaped bulbs that require no enclosures for weather-proofing.

two floodlights, cost only about $1 more, and three-bulb models are still under $7. The bulbs themselves cost an average of $2.29 each.

Many floodlights are mounted on garages with thin-walled pipe extensions for maximum illumination. They are most often used by homeowners who come home late at night and want the protection of plenty of light for the walk from the garage to the house. Floodlights are seldom left on all night because they create so much light that they can be disturbing to the neighbors.

Some homeowners also mount floodlights high up on the corners of the house itself, to cover sides of the building not illuminated by street lights. When such floodlights are turned on from inside at the sound of suspicious noises in the yard, they practically light up the whole neighborhood. Because the biggest part of a floodlight consists of the clear bulb itself, the installation of floodlights is unobtrusive, and many visitors do not even notice them when they are not turned on.

Ordinary incandescent light bulbs, screwed into the middle of a metal reflector like the ones used by warehouses over their doors, cost a lot less than floodlights. The fixtures

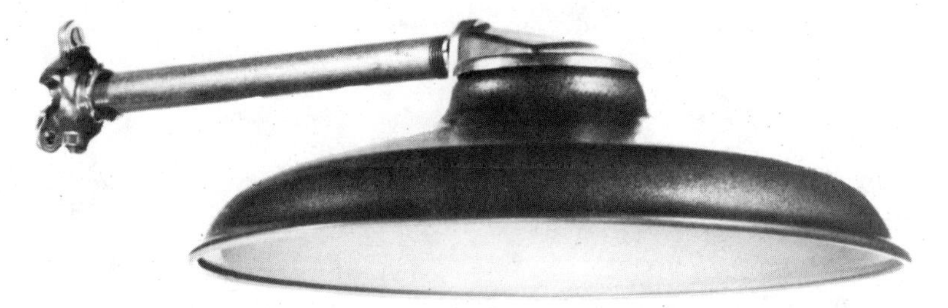

Ordinary reflectors use standard incandescent light bulbs, for the least expensive of all types of outdoor security lighting.

are less than pretty for residential use, but they don't cost much, either. A typical weather-resistant one, with an aluminum reflector fourteen inches in diameter, costs about $5. Steel models that are twelve inches in diameter cost only about $3.50, most of which is represented in the cost of the $2 reflector.

Mercury-vapor lights represent considerable initial investment, but provide much more light per watt used, and last many times longer than conventional lights.

These fixtures can usually take anything from 15-watt bulbs (costing about twenty-nine cents up to 100-watt sizes) to 150-watt incandescent bulbs (forty-five cents each). Turning them on and off repeatedly, however, can cost more than letting them burn continuously. The power surge each time they are turned on burns out bulbs rapidly, and can even use more electricity than allowing them to burn all the time.

Mercury-Vapor Lights

The homeowner seriously considering yard illumination should check into mercury-vapor lighting systems. The reason so many shopping centers and other public areas are illuminated with mercury-vapor bulbs is because they last so many times longer than the conventional incandescent light bulbs, and give so much more light per watt besides.

The initial investment is considerable, though. A typical yard light designed for residential use, in a post-mounted lantern with white opaque shatter-resistant panels and a 50-watt bulb, will cost around $70—not including the post. The replacement bulb alone is $9.

The automatic floodlight made by Sylvania, with a 175-watt mercury plasma bulb that turns on at dusk and off at dawn, retails for $65. That price includes a break-resistant plastic light refractor and a twenty-two-inch mounting bracket. Without the photoelectric cell, it's $10 less. The replacement bulb costs $10.

Wiring Systems

A photoelectric night switch, which turns yard lights on and off automatically at dusk and dawn respectively, is available for any type of light as an accessory for about $10. Good ones are not affected by car lights, having a built-in time delay to prevent accidental turn-offs. The system will, of course, adapt itself to seasonal time changes without any adjustments by the homeowner.

For another dollar or so, the automatic night switch can also incorporate a three-wire, grounded convenience outlet that the homeowner can use to run an electric lawn mower, hedge trimmer, barbeque grill, or any other outdoor appliance. Any such outlet should always have a weather-proof snap-down cover, to prevent shorting of the circuit in case of rain.

Manually operated yard lights should have two-way light switches so the homeowner can turn them on and off either from inside the house or from an outside switch, as possibly from inside the garage. If the homeowner has a burglar alarm as well as outside lighting, the two systems should definitely be hooked up together so that tripping the alarm will simultaneously light up the yard.

Many electrical contractors who install exterior security lighting advocate the use of underground cable to prevent burglars from cutting the power source. An underground cable is a good idea from the standpoint of aesthetics, but it is meaningless as a security measure when all a burglar has to do is lift a panel and unscrew the light bulb, or throw a rock at it. However, his putting out the security light should serve to alert the neighbors. A good electrical contractor should be concerned only with the safety and durability of his installation.

Interior Security Lighting

The best light to keep on at all times inside the house is the one inside the entrance at the front door. This will not only deter many a potential burglar from attempting an intrusion, but will also put *him* in the light, not you, if he does get in. Snapping on your bedroom light can be a bad idea if you hear an intruder in the house; it will only serve to illuminate you while the burglar remains invisible in the darkness.

A switch that turns on a light behind the burglar can be an excellent security device. This will attract his attention to a location that has nothing to do with you—and you can holler "Freeze!" when his back is turned to you. Furthermore, *he* will be the target in the light, not you.

Once a burglar is caught, have someone turn on *all* the lights in the house. Bright light usually immobilizes the average burglar, and you can hold him at bay until the police arrive.

Automatic Timers

The use of timers by security-conscious homeowners while they are out of town, to turn lights on and off automatically to make an unoccupied house look lived in, is almost universal (see Chapter 2 for tips for vacationers). But these clockwork timers also serve admirably for people who are just out for an evening. The cost of these devices is relatively low, and some people use a whole houseful of timers to make the place look occupied when nobody is home.

Plug-in timers, which can be used in any wall outlet, cost as little as five dollars apiece. The best-known manufacturer is Intermatic, Incorporated, Spring Grove, Illinois, whose president, James C. Miller, has a background that qualifies him as a guy who probably knows more about burglary psychology than most law enforcement officers. Miller is a positive believer in light as the most effective crime deterrent and is even trying to establish a National Burglary Protection Week, through his public relations representative Andy Tobin (whose house was burglarized in the summer of 1974 after the publicity given to his tour of anti-burglary speeches, when everybody knew he would be out of town). Plug-in timers, as well as more elaborate types, are sold by most hardware dealers, including mail order catalog houses such as Wards and Sears. The householder simply plugs the light into the current-connected timer, sets the timer for when he wants it to go on, and makes a second setting for when he wants it to turn off.

Plug-in timers can turn lights and appliances off and on to give the house a lived-in look while the residents are out.

A homeowner with simple timers throughout the house can set up a "lived in" pattern any way he likes. For example, when nobody is home, the timers could be set to produce the following pattern:

7:00 P.M.—Living room light goes on (dusk in summertime)
7:15 P.M.—Kitchen light and radio go on (doing the dishes)
7:45 P.M.—Kitchen goes dark (housework finished)
8:00 P.M.—Television set goes on
8:30 P.M.—Bathroom light goes on for fifteen minutes
9:00 P.M.—Children's room lights go on (preparing for bed)

Simple plug-in electric timers, such as the ones sold by the mail order companies like Wards and Sears, cost as little as five or six dollars apiece and can be used economically in series.

9:30 P.M.—Children's room lights go off (children in bed)
10:00 P.M.—Kitchen light goes on (snack time)
10:15 P.M.—Kitchen light goes off
11:00 P.M.—Bathroom light goes on
11:05 P.M.—Master bedroom light goes on (Mother preparing for bed)
11:15 P.M.—Master bedroom radio goes on (Mother listening while doing her nails)

11:30 P.M.—Master bedroom light, radio, go off (Mother goes to bed)
11:35 P.M.—Television set goes off (late news is over)
11:45 P.M.—Living room lights go off (ashtrays are emptied)
12:15 A.M.—Bathroom light goes off
12:16 A.M.—Master bedroom light goes on (Father preparing for bed)
12:25 A.M.—Master bedroom light goes off (Father goes to bed)
1:30 A.M.—Bathroom light goes on for five minutes

That entire staging is done with seven individual timers. They can be programmed for a twenty-four hour sequence, with the lights going on and off any desired number of times during the cycle. The homeowner who does not want that elaborate a system can still make a house look lived in with a single timer in the bathroom or kitchen. Turning on the light for five or ten minutes at a time every couple of hours or so leads an observer to think that the householder is not only home but also restless and having trouble getting to sleep.

Ultrasonics

Notable among low-cost motion detectors is the little ultrasonic unit made by the 3M Company, which retails for less than $100 in some areas. It will turn on a lamp (or any other appliance such as a radio) connected to the unit if anybody gets near it.

Measuring only three inches tall, six inches long and five inches wide, the solid-state, thirty-six-transistor plug-in unit steadily emits silent, harmless, high-frequency sound waves in a directional cone covering 300 square feet of floor space in front of it. Any movement within that protected area will activate the switch.

The 3M motion detector can also be set to turn on a

Motion detectors, like the $100 model made by 3M, can be used to turn on lights, radio, or any other electrical appliances at sign of intrusion.

small internal burglar alarm, although it's not much good as an audible alarm because it doesn't make enough noise to be heard in other rooms of the house if the doors are closed. But used strictly for turning on lights to startle an intruder, it can be particularly effective if hooked up with a strobe light, which makes many people think that their picture is being taken. Small strobe lights (seven inches tall), with solid-state circuitry and variable speed controls to regulate the frequency of the sun-bright flashes, are available for less than ten dollars.

10

Perimeter Barriers

"Fences are not built to keep burglars out, but to keep dogs in. Why take a chance when there are plenty of houses without fencing?"—Inmate, House of Correction, Detroit, Michigan.

The homeowner does not need a moat around his house to protect the property. But even a fourteen-inch picket fence, maybe to grow flowers against, is a good psychological barrier for unauthorized persons. People who would feel perfectly free to wander onto somebody's property, if its borderlines were unmarked, think twice about opening a private gate or going over a fence. Trespassing is itself a misdemeanor, especially when the property line is delineated with a barrier, in most communities.

No residential fence, wall, or hedge will keep out a professional thief intent on burglarizing the house, but it can deter plenty of "thieves of opportunity" who pilfer what they can when they can see anything of value. Even a hedge of bushes across the front of the property, without a gate of any kind at the walk leading to the door, will tend to keep trespassers out, merely by serving notice that it encloses private property.

Hedges

The ordinary privet (an ornamental shrub of the olive family) hedge is the most widely used boundary marker because this hardy plant grows almost anywhere and requires a minimum of care. Thickly growing hedges are often allowed to grow quite tall for the sake of privacy in the backyard, but they should be kept clipped short at the front of the house so they won't cut off the view of the house from the street. Houses built on corner lots should never have hedges more than a couple of feet high around the corner so that motorists can have a clear view of any traffic approaching from the intersecting street.

Hedges facing the street should not be so thick or so tall as to provide hiding places for burglars. For this and other reasons, an eighteen-inch privet hedge does its job as well as a four-foot one; a small boy retrieving a baseball from the yard can jump over a knee-high hedge, but he will break through a high privet hedge, sometimes destructively, because there are no thorns to discourage him.

Hedging that does have thorns includes holly, barberry, and multiflora rose bushes. Thorny bushes should all be kept low, lest small children run into thorns at eye level. If they just get scratched up a little from playing where they have no business to be playing, all well and good; they get what they deserve. But nobody wants to put a small child's eye out just for getting into mischief.

A quite safe, as well as quite effective, barrier is a wide planting of thorny bushes behind a three- or four-foot fence. Although the fence is not hard to climb, anybody who tries it faces the prospect of dropping down into a tangled mass of thorns. If the bushes are planted about a foot apart, there's no way to get through them without getting clothing snagged on the thorns, not to mention scratching a little skin. Such bushes cost about twice as much as ordinary privet hedging—thirty or forty cents apiece in dozen lots from most nurserymen. They are also harder to take care of and require a great deal of watering.

Wood Fencing

An ordinary picket fence is a better choice for a front

yard than the shoulder-to-shoulder stake fences becoming so popular as "stockade fencing," because the spaces between the pickets provide visibility of the house from the street while keeping out even a small cat. The spaces between each upright picket should be as wide as the pickets themselves but should be too small for mischievous children to get their feet into.

Sharpening each picket to a point at the top serves no practical purpose and may injure some child trying to climb the fence. Stockade fences are commonly pointed at the tops of the stakes and are almost always built of cedar for the sake of durability. Anybody who can afford cedar at all can afford to have it installed, too, instead of putting it in on a do-it-yourself basis. A six-foot stockade fence made of White Michigan Cedar will cost the homeowner about seven dollars per linear foot, or about eight dollars a foot for Western Red.

A cross-rail fence, as often used for Colonial-styled homes (and on horse farms), is better looking than a picket fence, and it does provide street visibility but not the tightness of security (children can crawl through it as well as over it). Rail fences built with horizontal rails rather than with crossed members are not widely used because the bottom rail must necessarily be fastened to its supporting posts at a considerable height from the ground to keep it from rotting.

Solid board fences, like the one Tom Sawyer white-washed for his Aunt Polly, are used primarily for alley fencing where privacy for the backyard is desirable. Like all wood fences made of pine (the price of cedar is ridiculous), they should be painted every year. Using 1 x 6-inch boards for the uprights will require more nailing than if 1 x 8s are used, but there's less chance of cupping and warping if the narrower boards are used.

There are certain requirements the homeowner must bear in mind no matter which of the above types of wooden fencing he chooses to use. A wooden fence should have a

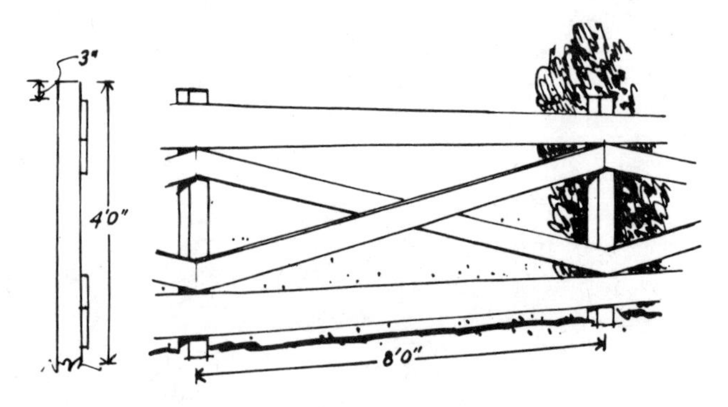

Cross-rail fences make good use of a minimal lumber supply, but are harder to build than many other types of wood fencing.

post sunk into the ground every six or eight feet apart for proper support. The bottoms of fence posts *must* be heavily coated with creosote to protect them against wood-boring insects and rot. Unless the posts are anchored in concrete, they should be sunk three feet into the ground to give the 2 x 4s supporting the fencing members a solid base.

When considering building a wooden fence, bear in mind the fact that lumber prices were among the first to indicate that the law of supply and demand was finally starting to work in the summer of 1974. When the new interest rates for mortgage money gave so many potential buyers pause, the price of No. 3 1 x 6-inch pine boards dropped from 23¢ per running foot in June to 14¢ in August. The necessary 2 x 4s are still 20¢ a foot, though, and seven-foot posts five inches in diameter are still $1.75 apiece.

Redwood, often used for privacy fencing by home-owners in new housing tracts, is not all that much more impervious to weathering than pine. However, it costs about the same as No. 2 pine (fewer knots than in commercial No. 3 pine), and it is often built in a "basket-weave" pattern in the interests of looking different.

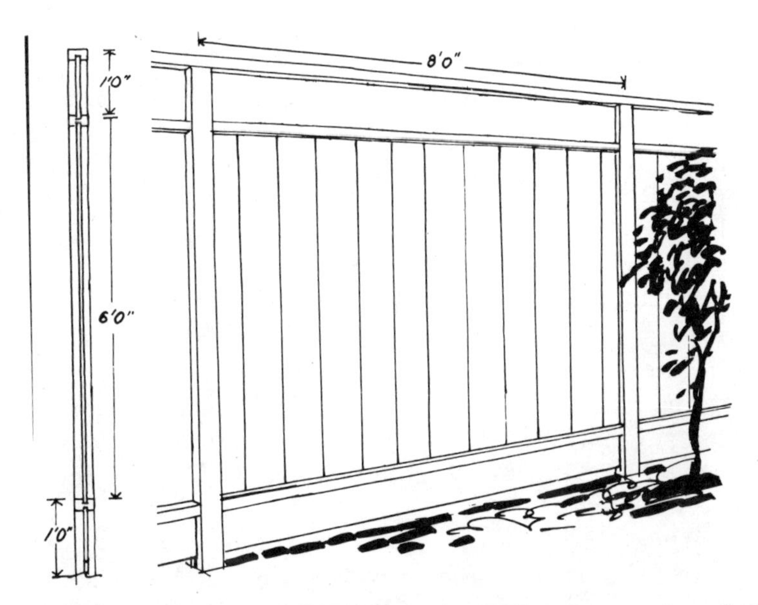

Privacy fencing high enough to obstruct the view into the owner's yard are fine as protection against nosey neighbors, but burglars like them because they can work behind them without being seen.

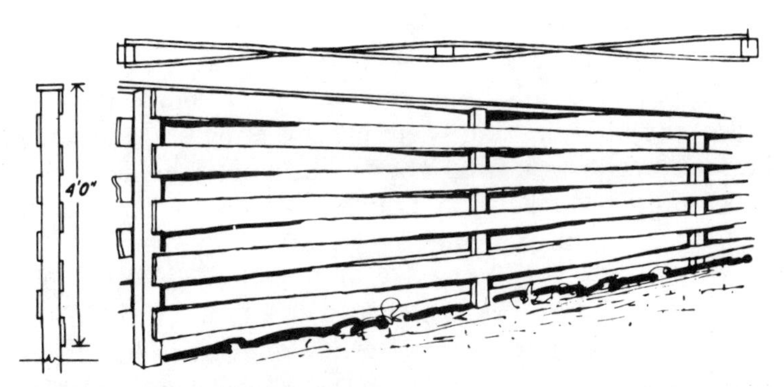

Basketweave fencing's big advantage is that it looks the same on both sides, and has a lot of appeal for homeowners who like a little extra flair.

Chain Link Fencing

The most widely used metal fencing is the chain link fence—which many people generically call a Cyclone Fence, although that name belongs to U.S. Steel. The economy grade is woven of 11½-gauge galvanized wire with 2¼ x 2¼-inch diamond-shaped mesh—just the right size for a child to get the toe of his tennis shoe into. A 48-inch height costs around $1 per linear foot, including standard 1-5/8" line posts ten feet apart, 1-3/8" top rails, and aluminum tie wires, components that account for about half the cost of a complete fencing outfit. Corner posts, gate posts, and gates are extra, from $7 to $8 for corner posts, $5 to $7 for gate posts, and $20 or so for a single walk gate.

A heavier 11-gauge grade, with more closely woven wire in a 2 x 2-inch mesh, provides considerably stronger fencing, and costs only about 10 percent more in a 48-inch height. The same closer mesh, but fabricated of heavy-duty 9-gauge wire, costs almost half again as much as the economy grade. For another ten cents or so a foot, the 9-gauge 2 x 2-inch chain link fencing is also available with a vinyl coating that will not rust, chip, peel, or fade.

Standard chain link fencing is made in 36-, 42-, 48-, 60-, and 72-inch heights. A six-foot fence costs only about 50 percent more than a three-foot fence, not twice as much, because of the lower cost per foot for the posts and top rails.

At one time, all chain link fencing was made with the ends of the stiff wire simply cut off at the top and bottom of the fence. The resulting short metal spikes make such a fence quite hazardous to crawl under, and woe betide anybody who slips while trying to climb over it. The protruding cut-off wires were sometimes twisted together to form a two-point barb. However, this type of chain link fence resulted in a horrendous number of gashed hands, cut faces, and punctured groins—especially horrendous considering that these fences were commonly used around

Chain link fencing, often known as Cyclone fencing although that's a trademarked name owned by U.S. Steel, is more durable and provides better protection than wood fencing.

playgrounds where children too lazy to walk to a gate frequently got themselves hung up on top of the fence.

Most chain link fencing today is made with the ends of the heavy wire paired and looped together flat, so as not to stick out but to form a smooth edge that is not likely to cut anybody climbing over it. Dangerous wire fencing, including standard barbed wire, traps mostly the relatively innocent; it will not stop a professional thief (he just throws a jacket or a tarp over the top of the fence and *then* climbs over it).

Other Types of Metal Fencing

Chicken wire won't contain chickens, much less burglars, but it's cheap. Technically known as wire netting, the hexagonal mesh 20-gauge steel fencing costs only about ten dollars for a 150-foot roll in a strip two feet wide. That figures out to be less than seven cents per linear foot. It is made in 50- and 150-foot rolls of strips from twelve to seventy-two inches in width.

What most chicken growers actually use for fencing, rather than the hexagonal so-called chicken wire, is sturdier poultry fencing made of 15½-gauge rectangular mesh with a forty-eight-inch width having twenty line wires (spaced only an inch apart at the bottom to keep chicks in the yard), with 12½-gauge wires at top and bottom for extra strength. It costs about ten cents per linear foot. For another two cents a foot, this type of wire is made for field fencing with just ten line wires and with the vertical stay wires spaced six inches apart, both made of 12½-gauge steel but with heavy 10-gauge wires at the top and bottom. There's cheaper heavy-grade field fencing, too, with only six line wires and the stays spaced a foot apart, which costs even less than chichen wire netting.

Still another type of metal fencing sold in rolls of various widths is welded wire mesh. This is often used for building dog runs or for separating a part of a yard for small children to play in. With a 2 x 4-inch rectangular mesh made of 11-gauge welded wire in a forty-eight-inch width, it costs about forty cents per linear foot.

Of the five standard mesh sizes for welded wire fencing, the most expensive is the ½ x ½-inch mesh at over fifty cents per foot. This is more than the cost of chain link fencing when it's bought in rolls.

The cost advantage in using any of these types of metal mesh fencing is in their simplicity of installation. They do not require the use of top rails, and the meshed wire is simply stretched between low-cost angle-iron posts. Such

Fencing made of welded wire mesh can provide low-cost enclosures in grades from "chicken wire" to heavy 10-gauge steel.

posts, punched with self-fastening lugs, can easily be hammered into the ground, and they cost only about $1.25 each for a four-foot fence.

Not much seen anymore is the ornamental "wire picket" fencing once so widely used to enclose suburban yards. The deeply crimped 11-gauge wire is looped in a top arch between each pair of uprights in the "single picket" style, and costs about thirty-five cents per linear foot for a forty-eight-inch height. The "double picket" style, which has arches and double vertical wiring in the bottom half, is about forty-five cents a foot; with heavier 9½-gauge wire, it costs over fiftv cents a foot.

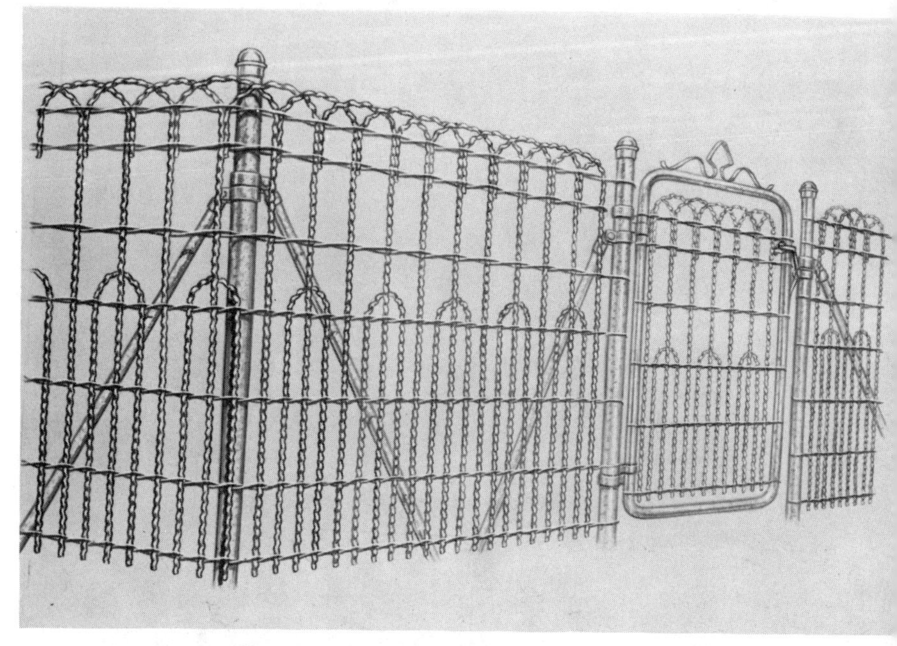

Wire-picket fencing provides good protection and is economical because it does not require top or bottom rails.

The 1-5/8-inch line posts used for wire picket fencing, complete with post caps and tie wires, are approximately $2.50 each. Heavy two-inch galvanized steel posts, with one steel brace for use as end or gate posts, are $6.50 each, or about $9 if two braces are used as corner posts. Single walk gates, with a one-inch galvanized steel frame, are about $12.

Ordinary strung wire, costing less than two cents a foot in 19-gauge galvanized steel, is most often used with a low-cost charger for electrified fencing. A solid-state charger, which intermittently delivers 3,800 volts of electricity and has a frequency switch to increase the number of shocks, operates in any weather and costs under $45 for a U.L.-approved model. Some people prefer to electrify barbed wire, which can cost less than five cents a foot for 18-gauge steel with two-point barbs every four inches, or as much as twelve cents a foot for 12½-gauge steel with four-point barbs. A supply of two-cent insulators for the posts will complete the system.

The most expensive residential fencing is made of wrought iron. A plain forty-eight-inch wrought iron fence, with uprights four inches apart, now costs over $20 per linear foot to have installed (it is a very difficult job for a do-it-yourselfer, although few homeowners who can afford to consider wrought iron fencing are very much concerned about cutting costs by supplying their own labor). Elaborate gates, eagle-crested corner posts, and scrollwork can take the installation of wrought iron fencing out of the field of journeyman craftsmanship and into the field of art, where the sky is the limit for costs. Wrought iron also requires steady maintenance to prevent rusting.

Fencing made of aluminum bars and rails looks like wrought iron when it is painted but costs only about half as much for plain running lengths. Aluminum won't deteriorate like iron, it is easier to install, and it is actually stronger as well. It has done to wrought iron what the Chevrolet did to the Packard.

Walls

Masonry walls do not have a place in ordinary residential use. They are picturesque in New England, where the colonists clearing their fields didn't have anything else to do with the rocks except build fences out of them, or in Ireland, where the people didn't have enough money to buy lumber. But a solid rock or brick wall cuts off the visibility needed for good security, interferes with breezes, and makes it hard to grow plantings alongside it.

Today's cost of constructing one is all but prohibitive for anybody with less than the biggest of estates. The cost of various types of stone, brickwork, and labor are all wildly variable, depending on the height, thickness, and length of the construction. Construction of the Great Wall of China was possible only because of coolie labor, but with the advent of American labor unions, building a property-enclosing wall can cost as much as the house itself.

Stucco is sometimes used for wall construction, but this cement, plastered to a mesh base, is seldom durable enough for climates that have freezing temperatures. Stucco walls are sometimes seen around imposing older residences in Florida and California, but not many new ones. The type of person who wants to impress his neighbors with the grandeur of a wall around his property but who uses stucco to save money is a little out of style these days.

Building brick walls around his property was one of the many hobbies of Winston Churchill, but he did it for the satisfaction of working with the tools and materials, not for any practical purpose. He liked what he considered the aesthetics of the finished product, too. Aesthetics, of course, are in the eye of the beholder. Frank Lloyd Wright once designed a house for a movie mogul who, a few months later, built a brick wall between his new home and the one next door. This so incensed Frank Lloyd Wright, who considered the wall a debasement of his conception of what the total environmental landscape should look like, that he hired a crew of workmen and demolished the wall the first day the movie maker was out of town.

Spike-studded walls, or walls with broken bottles cemented along the top, are illegal in most areas. In fact, so is barbed wire, electrified fencing, or any other kind of barrier that could hurt an innocent person (how innocent anybody is who tries to climb over such a barrier remains for the courts to decide). The primary reason for making such walls a legal liability is the hazard of mischievous kids who could hurt themselves.

Various municipal codes also restrict the heights of

walls, fencing, or hedges for a supposedly more pleasing atmosphere in the neighborhood. Others have regulations regarding the wall's distance from the property line or the sidewalk to provide a more spacious appearance. Some communities have ordinances against the use of metal fencing in the interests of conformity, sometimes not even allowing fencing at the front of the house at all lest visitors (and/or potential real estate buyers in the developing housing tract) get the impression that residents have been scared into being security-conscious.

Any homeowner considering the installation of a perimeter barrier of any kind should always check out his plans and specifications with the city hall before buying anything. In some cases he will even have to get a building permit. The red tape sometimes involved can be unbelievable, and the homeowner is well advised to get the help of his representative in the city government (alderman, councilman, or whoever it may be).

11

Window Gratings

"I don't want to have anything more to do with any place of any kind that has bars on the windows"—Convict, State Penitentiary, Elmira, New York.

America is the only civilized country in the world where most well-off homeowners do not have protective gratings on the ground floor windows. Romantics would have us believe that this grillwork is to keep out impassioned caballeros from the senoritas when the duennas are not watching (or to keep in impassioned senoritas in the Rita Hayworth movies). But the mundane truth of the matter is that window grills are to keep out burglars.

In countries where most people have almost nothing, the people who do have property of value go to great lengths to protect it. In this country, where almost everybody owns property worth stealing, the good things in life are so much taken for granted that owners are more apt to be lax in taking such security measures.

Small-Pane Windows

Originally, windows were made with a number of small panes because making glass in large sheets was impossible. Today, most small-pane windows are used to enhance the appearance of a Colonial house. They have a practical purpose, too, in that replacing a small pane costs maybe a sixth the cost of replacing a whole window if it gets broken.

Their limitations are in the higher initial cost, the work it takes to keep them clean, and the really tough job of painting and maintenance. But they have a security function, too. Breaking out a single pane of glass in a small-pane window does not provide enough room for an intruder to crawl through.

There are plenty of other ways for a burglar to get in through a small-paned window (see Chapter 2), but the appearance of a multiplicity of small apertures is enough to discourage many prowlers.

People who like the looks of small-pane windows, but who do not want to replace their existing double-sash windows with completely new windows, can still get the same effect with wooden strip bars. Fastened to the outside of a regular double-sash window, the horizontal and vertical pane bars simulate real ones quite effectively. The wooden moulding of which pane bars is made is available at any lumber yard for a few cents a foot.

Cutting and fitting the molding can be tricky for a do-it-yourselfer, especialy where the vertical and horizontal bars meet. Precut kits are now available, usually made of white vinyl for easy maintenance, to fit most windows. A kit for a double-sash window with a standard 26 x 30-inch pane in each half, costs less than four dollars for two horizontal bars and four vertical bars (to simulate six small panes in each half). Kits for smaller windows cost even less, and the ones for bigger sizes, with six vertical bars, are still only around five dollars.

Shutters

The Colonists used to close their window shutters to

protect the then-valuable window glass during hail storms, to provide insulation against cold or the heat of the sun, to give themselves privacy (instead of shades, which they didn't have), and sometimes to repel Indian attacks. Most window shutters today are strictly decorative, often merely being nailed onto the house at the sides of the windows. Functional hinged shutters, though, can present a formidable-looking barrier to would-be intruders when closed and hooked shut, especially when there's a light on behind them.

Good wooden shutters have louvers for ventilation when closed; these louvers should be mortised into the frame so they can't be pulled out. The shutter at each side of the window necessarily has to be half the width of the window, and shutters sixteen inches wide cost about twelve dollars a pair for a fifty-one-inch height. For various heights, the homeowner can figure about twelve cents per inch per side for the sixteen-inch widths, although narrower widths are available in standard inventories, too. A set of relatively simple hardware, including four hinges and a pair of S-shaped iron hold-backs, is usually under three dollars complete.

Shutters enclosing windows more than thirty-two inches wide are most often folding shutters, so that they won't take up so much room against the side of the house when not in functional use. Folding shutters cost about half again as much as solid shutters. Many such shutters are often used inside the window, where they are frequently made with a vertical tilt bar to adjust the angle of the movable louvers to control light and privacy, comparable to Venetian blinds.

Regardless of the type, all slatted wooden shutters are the very devil to take care of in year-after-year maintenance. They are subject to warping, cracking, rot, and mildew. Repainting them is a chore no homeowner looks forward to. Even the hinges and hold-backs on exterior

Window shutters intended for security should have ventilation louvers mortised into the frames so they can't be pulled out.

shutters need frequent attention to keep them from rusting, lest rain dripping from them mar the finish of the house. Technology has now come up with—what else?—plastic shutters, which are practically maintenance-free.

Shutters made of heavy, tough, high-impact polystyrene are molded in one-piece construction, including the ventilating louvers. Most polystyrenes are made with wood graining molded into the surface for the look and feel of wood, prefinished in black or white. The standard width is fourteen inches per side, and in a fifty-one-inch height they cost about $16 a pair.

Lightweight cheapies in the plastic shutter field have solid molded panels that only look as though they are louvered. That means no ventilation, which makes them less than ideal for use in anything but a purely decorative application. Even there, the lack of air circulation behind them retards drying of the wall surface to which they are affixed.

A pair of these shutters in fifteen-inch width costs about ten dollars for a fifty-one-inch height. Stock heights range from less than two feet (six dollars a pair) to almost eight feet (fifteen dollars a pair). Many buyers also add needless S-shaped "hold-backs" made of styrene with the supposed appearance of wrought iron, costing about one dollar a pair.

Grillwork

The main reason scrolled aluminum grillwork is so often used on combination screen doors is for security, not just for decoration. The same type of grillwork is being used increasingly for windows fronting the house. Windows opening onto the front porch, in particular, are especially vulnerable to illicit entry unless protected with gratings.

The biggest disadvantage of permanent window gratings is that people can't use the windows as an emergency exit in case of fire. A window right next to the door, though, seldom needs to be used as an exit, and porch windows can be protected with the most immovable of fixtures.

Aluminum grillwork on a porch window matching the grillwork on the door is not unattractive, and aluminum has the advantage of being rustproof, easy to work with, and cheap. Depending on complexity of the design, an aluminum window grill for the average-sized window with no opening big enough for entry costs from five to ten dollars. The main disadvantage of aluminum is that it is comparatively easy for a burglar to cut through.

A strong window grill fabricated of ornamental iron will cost two to three times as much as aluminum, custom-made and installed. The biggest cost of a wrought iron grill is in the labor involved, which the average do-it-yourselfer is ill-equipped to do. Ironwork can be as intricate as the homeowner's pocketbook allows (see Chapter 10). It literally can be art.

Permanent window grills, of any type, could be installed on possible exit windows such as those opening onto fire escapes, but *only* if they can be opened from the inside. A key for the inside lock should be instantly available in case anybody ever needs it in a hurry. The gratings cemented into concrete, as in the classic hacienda-type grills (sometimes made of carved wooden bars or heavy lattice-work) ignore fire safety precautions at the risk of crisping the occupants.

Utilitarian Window Guards

When a burglar is brought to the police station, his first taste of being locked up is often in a detention cage while the police check him out, not behind bars. These enclosures built against a wall of the squad room are made of woven wire mesh only a little heavier than standard chain link fencing. This mesh is plenty strong to keep burglars out of a house, too, when it is used in a steel-

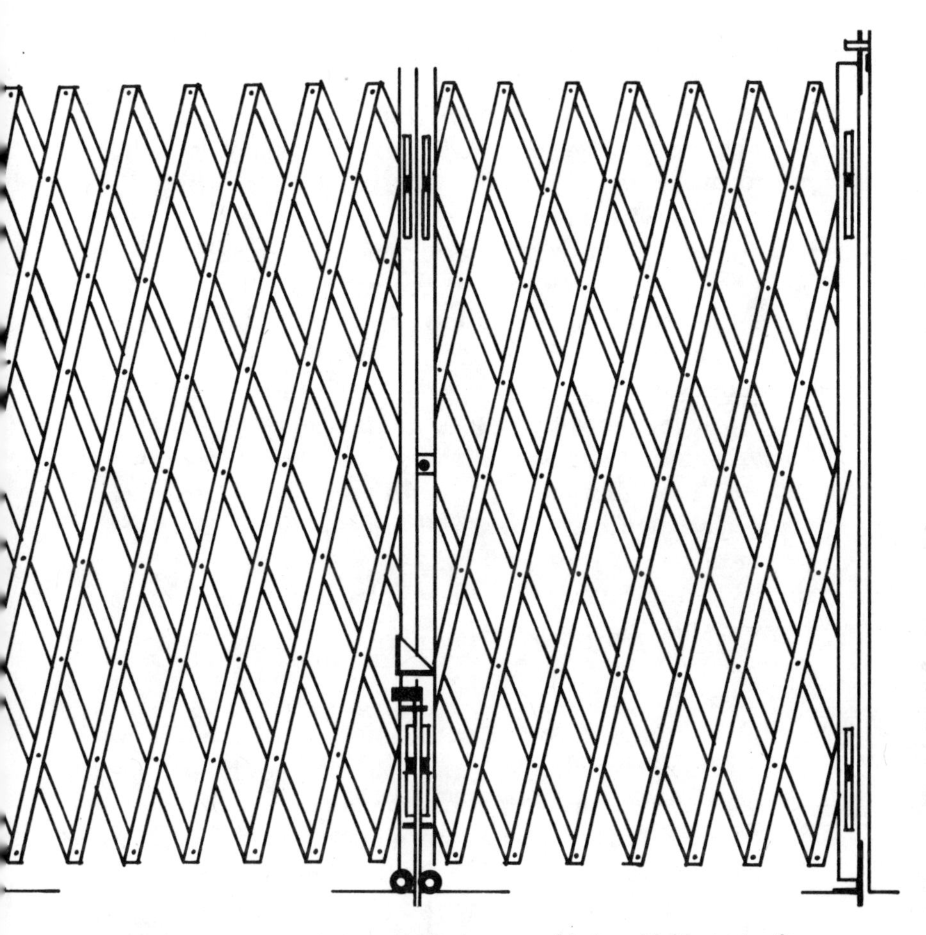

Folding gates covering doors and windows provide formidable protection when closed, but their security is no better than the quality of the locks used to lock them up.

framed window guard. The heavy-gauge welded mesh costs less than twenty cents per square foot, but it is too unsightly to be used on windows other than in basements or garages.

Ordinary meshed wire fencing (see Chapter 10) can also be cut to size and used in window guards at minimal expense. The smaller the size of the mesh, down to ½ x ½-

inch for welded wire, the more pieces a burglar has to cut through to effect an entry. Small mesh sizes utilize thin-gauge wire, however, which can be snipped with electricians' pliers. Anybody trying to cut through a window guard made of 11½-gauge wire, or heavier, will have to carry a big wire cutter.

Window guards should be removable or hinged at the top or a side, if only so that the windows can be washed from the outside, regardless of their possible use as emergency exits. Some window guards are made of fold-in crisscross steel slats, of the same design as wooden "baby gates," or as increasingly seen in ghetto areas to cover the entire fronts of stores at night to foil the smash-and-grab thieves. Their inside locks should be shrouded to prevent tampering, with the keys within easy reach of anybody trying to get out (but out of reach of anybody trying to get in). Window guards should also have plugged hinges to prevent their pins from being pushed out.

Iron Bars

A window protected with iron bars is impossible for the average burglar to get into. Impossible for the occupants to get out of, too, but that's a matter of priorities.

Some window guards are made of iron bars, but a swing-away window guard fabricated of heat-treated iron bars offers no more security than one made of "detention cage" wire mesh. The only real reason for using iron bars is to prevent anybody from going through a protected window from either side. Barred windows are most often built with both tops and bottoms of the bars cemented into the masonry.

They can be cut, but only with the expenditure of a lot of time and energy with hacksaw blades. When the bars are three inches apart (or closer), entry requires the cutting of more than one bar, both top and bottom, and few burglars are willing to do that much work for a mere residential hit.

Not even the small-pane windows utilizing steel between the panes instead of wood, as in factories and jails that also incorporate wire mesh embedded in the glass, are as effective as iron bars on a window for security. Putting tempered 3/4-inch iron bars on a window during construction of a house will add $25 to $30 to the cost of the window. But it will keep the rascals out as nothing else will.

12

Safes

"A Jimmy Valentine of the 1970s doesn't sandpaper his fingertips to work the combination lock of a wall safe, he uses a sledge hammer to knock the whole knob off"—Convict, Federal Penitentiary, Leavenworth, Kansas.

The average safe used by homeowners, to keep their valuables in, resembles a good burglar alarm in that it affords fire protection as well as theft protection. In fact it provides *better* protection against fire than it does against burglary. That's because insulated vaults designed to protect important papers and property against fire are the most commonly used residential safes, inasmuch as they cost less to install than a true, high-security money safe. Their construction is good enough to keep out average burglars—not the Alexander Mundy type of professional safecracker, but the 95 percent of all burglars who know practically nothing about getting into any kind of a safe equipped with a combination lock.

Portable Vaults

Some vaults do not afford as much burglary protection as an ordinary filing cabinet with a lock on it. A $60 four-drawer steel filing cabinet with a lock to challenge burglars

Portable safes are better for privacy than for burglary protection, especially when the safe is small enough to be carried out under the burglar's arm.

is too big to carry out easily, but some of the "security vaults" weigh as little as fifteen or twenty pounds, and can be carried out under the burglar's arm so he can open them up at home at his leisure.

A $5 or $10 fishing tacklebox or a toolbox with a padlock on it will serve as well as some of the small security vaults sold by manufacturers such as the Meilink Steel Safe Company, Toledo, Ohio. Their vaults are made with interiors as small as 4¾ x 4¼ x 10¼ inches, and cost around $30 for models that will protect their contents against fire up to half an hour at 1550° F. temperatures. Meilink vaults with similar pull-out drawers big enough to hold, say, confidential card files cost around $50.

Their letter vault, also with a filing-cabinet-type drawer, is around $100 for a 10 x 12 x 13½-inch interior, and it will withstand 1700° F. temperatures for one hour. With a top-opening lid, which makes it look like a picnic-type beer cooler, the one-hour Meilink letter vault in a bit smaller size is around $70. The buyer has a choice of key or combination lock.

Letter vaults are basically designed for fire protection rather than for burglary protection, and some can be opened with a can opener.

Smaller-size Meilink vaults with top-opening lids cost as little as $60 for one-hour furnace-tested models. Price competition retails top-lidded “security vaults” for as little as $10 for 14 x 4 x 9-inch models with asbestos lining between double steel walls, but without any official assurance of specific fire protection. Ward’s sells one such unit, big enough to hold file folders with a 14½ x 11 3/8 x 9½-inch interior, for $19.95.

Schwab Safe Company, Lafayette, Indiana, makes a

better grade chest-type top-lidded safe with nearly two inches of insulation in the double steel walls, and a door two-and-one-half inches thick for the one-hour protection against 1700° F. temperatures. A 16 x 10 x 8-inch model retails for about $70. It has either a key or combination lock, and the lid has a pull ring for easy opening. But it weighs less than sixty-five pounds—and conveniently has carrying handles on both ends.

Schwab also makes a little $70 safe resembling a safety deposit box. The 9½ x 9½-inch door has a combination lock, but the one-hour/1700° unit weighs just forty-five pounds, including the removable nineteen-inch drawer. No handles needed.

In-Floor Safes

Schwab's $67 floor safe, which has an exterior only ten inches deep and which can be sunk into a wooden floor, also has a Class C fire rating. This rating means that the safe protects contents for one hour in a furnace test at 1700°F., and the rating entitles the owner up to a 20 percent discount on his insurance premiums in some states. A safe with a Class B rating, as used in commercial applications, protects contents for two hours at 1850°F., with walls 3-7/8 inches thick, and can get owners up to a 30 percent insurance discount. The big Class A safe protects contents for four hours at 2000° F., has a multiple-step door six inches thick, and can reduce insurance costs by 40 percent.

Most in-floor safes are meant to be installed in basement floors during new construction and are primarily money safes without fire-protective insulation. Schwab's tube-style floor safes are not even UL-approved. They have an eight-inch inside diameter, and the $145 model is 11½ inches deep. The $200 version, 14½ inches deep, has a double compartment inside, with a separate locked door in the partition.

Schwab also makes a no-nonsense vertical floor safe

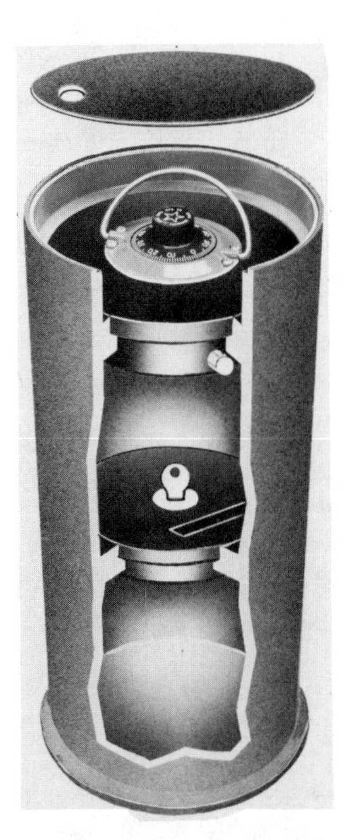

In-floor safes are designed to be buried in concrete during new construction. They provide good security, especially when equipped with an inner compartment that has its own separate lock, but they are inconvenient to use.

with electrically welded walls and bottom made of solid steel an inch thick. This high-security money safe has a round door 2¾ inches thick, and although the double-compartment interior only measures 8 x 8 x 18 inches, the price is $425. If the combination lock is attacked by tools, a relocking device automatically deadlocks three hardened steel bolts into the door. A similar version with comparable features is made by Mosler Safe Company, Hamilton, Ohio, with a 7 x 10 x 11-inch interior, for $358.

The Sentry line of safes made by the John D. Brush Company, Rochester, New York, includes an $85 tube-style floor safe that has a square 9 x 9-inch steel base welded to the bottom, which effectively prevents the 8 x 13-inch safe from ever being pulled up out of the concrete. The two-compartment version, with a choice of either a key lock or a combination lock for the floor-level door, is another $25.

Wall Safes

The classic place for espionage agents and millionaires to keep their valuables is in a wall safe hidden behind a picture or mirror. The homeowner can pay from $59.95, for a C-rated wall safe with an 11 x 9½ x 8 3/8-inch interior from Montgomery Ward, to over $2,000 for a torch- and tool-resistant model built by Mosler.

And that's a *small* Mosler, with a 14 x 14 x 12-inch interior. But it's built of a special steel alloy that resists attack by tungsten carbide and diamond core drills, the most effective of modern cutting tools. The inclusion of copper in the alloy also makes it highly resistant to attack by torch. In the event of frontal attack by sledge or punch, the relock device automatically deadlocks dual bolts in the door, and a thermal relock does the same in case of torch attack. The door is machined of special security steel 2 3/8 inches thick, and the combination lock (radiological-proof and manipulation-resistant) is capable of 1,000,000 different settings.

The homeowner who feels that he can get along with a Mosler that is not torch-resistant, but still wants one that's tool-resistant, can get the same size for $1,153. He can save another $300 or so by getting a model that is still secure against sledge attack, but not so resistant to cutting tools.

Not all Mosler wall safes are that expensive. Basically the same safe sold as an in-floor model (see above), but installed horizontally in a wall with flanges, is $413, complete with the relocking device to deadbolt the door in

Most wall safes are mounted to the wall studs . . . from which the entire installation is sometimes simply unscrewed by impatient burglars.

the event of mechanical or explosives attack.

Good safes often are made further burglar-resistant with the installation of cladding—a half ton of steel-reinforced concrete completely enclosing the sides and back of the safe to thwart bodily carry-off. A cladded safe also makes physical attack much more difficult. Better grade wall safes have round plug-type doors instead of the full-front rectangular doors that serve as the entire front wall on economy models when closed.

Most cheap wall safes are built to fit between wall studs with sixteen-inch centers. Each side of the safe has a metal ear that is simply screwed onto the 2 x 4 stud. The burglar does not need to open the safe; he can unscrew it from the studs and carry the whole thing home with him. When a wall safe is installed with non-removable screws (see Chapter 2), some burglars have been known just to saw off the 2 x 4 sections out of the studs so they can carry the safe away. If they don't wreck it getting it open, they can even sell it as part of the haul.

Ward's is by no means the only source for cheap wall safes. Schwab's $68 model is basically the same as their in-floor cheapie (see above), measuring 9½ x 14½ x 6 inches on the inside. Meilink's version, also with a C-rating for fire protection, is $86. Sentry also has two models at under $100, one with a bit of insulation that will protect the contents for an hour at 350° F. temperatures, and the other with a fire-resistant door but an uninsulated body, for installation in masonry walls.

A wall safe is like a magnet to burglars, and the underside of the carpet in front of it is a good place to plant a mat switch for the burglar alarm. A wall safe is also easy to wire up to give anybody touching it an electric shock; however, be advised that giving an intruder too much of a jolt is illegal in some states.

Standing Models

There's not much point in having a safe standing in a corner of the room if it is small enough and light enough in weight to carry out. A standing safe equipped with casters will even conveniently enable the burglar to wheel it out. Too many safes give a burglar just one more thing to steal.

Among the most secure residential safes are Schwab's line of TL-30 models, with the smaller ones designed to take cladding. The one measuring 21 3/8 x 21 3/8 x 19¼ inches inside weighs 442 pounds without cladding and costs $1,295,but it is constructed of armor plate steel, the same material a military tank is made of. Its tensile strength is 200,000 pounds per square inch, which is four times the tensile strength of ordinary steel. The smallest, with a 16 1/8 x 13 3/16 x 11¼-inch interior, is still $075, plus cladding. All TL-30s have the UL-approved relocking device for automatic deadlocking in case of tool attack. A

Floor-model safes should rest on the floor, not on rollers for the convenience of burglars in rolling them out.

cladded TL-30 will pay for itself in reduced insurance rates, with savings for broad form insurance of 40 to 60 percent.

But Schwab makes the little roll-arounds, too, with a C-rating for fire protection. The one with a partitioned 11 x 14 x 15-inch interior is $173; for a 13 x 17 x 15, $204. They weigh 175 and 238 pounds respectively—no big challenge for anybody with a furniture mover's dolly. The hefty 485-pound model from the Shaw-Walker Company, Muskegon, Michigan (at $510, plus $92 for a partitioned 31 x 18 x 21-inch interior) would take considerable effort to move, and the 9-inch-higher $610 model weighing in at 600 pounds even more. Both have C-ratings for fire protection, but their combination locks are somewhat of a joke for

A safe does not necessarily need to look unattractive, and can be housed in furniture-styled cabinetry for use even in living rooms.

anybody who knows anything about burglar's tools.

Meilink's biggest standing safe measures 17 x 12½ x 18½ inches inside, with a price tag of $208, but the C-rated safe weighs only 225 pounds. The company makes four other C-rated models in still smaller sizes of standing safes, down to and including a $69 model, with a 13 x 9 x 10-inch interior, which weighs a mere 75 pounds. At that size, the safe more properly should be considered a shelf model for the closet, rather than taking up floor space where the owner would have to get down on his hands and knees to use it. Schwab has a similar "safe," 11 x 13 x 13 inches inside, for $102, which weighs 90 pounds.

The big mail order companies such as Sears Roebuck and Montgomery Ward are in there pitching with C-rated roll-arounds, too. Ward's biggest, at $149.95, is 12½ x 15 x 18½ inches inside, and weighs 215 pounds. Their best-selling safe is the $129.95, 185-pound model with a 13½ x

16 7/8 x 12-inch interior, and their lowest-priced roll-around is the $89.95 model weighing 145 pounds, with a 10 x 13 x 9-inch interior.

The best-*looking* standing safes, if not the safest, are Brush's Sentry models. Instead of having starkly squared corners, which can be painful if bumped into, they have rounded corners—and are usually contained in furniture-styled cabinets supplied by Brush. The standard S-3 model, with a 15 x 12 x 13-inch interior, sells for $120 but is only $60 more delivered in a handsome cabinet with a choice of Traditional, Contemporary, or Mosaic styling, for a total weight of 205 pounds.

The bigger S-8 model, 6½ inches deeper, is good-looking enough in its own right so that this $165 unit does not need a cabinet. A Cadet model, measuring 15 x 12 x 9½ inches inside, is a flat $100. A key lock (more dependable on these small safes) can be substituted for the so-so combination lock on all models.

The lock on a safe engineered primarily for fire safety will stop common pilferage, but it cannot be taken too seriously for high-security burglary protection. However, burglary protection is the only reason for which the average homeowner needs to invest in any kind of safe. If the homeowner just wants to protect important papers (including money) from fire, he will find that he usually already has a good fire-proof vault—his food freezer. The insulation in the walls of a deep-freezer is better than most safes because it has to keep working to fight away heat *all* the time. Many bigger freezers, from 6½ cubic feet and larger, have locks on them, too. The moral: The average homeowner who has read this chapter can forget about it if he has a lockable freezer in the house.

13

Vehicle Protection

"I can hot-wire a car in less time than it takes some car owners to get the engine started with an ignition key"—Convict, State Penitentiary, Joliet, Illinois.

Jumping the ignition in a vehicle is so easy that no real expertise is required. The ignition is like any other electrical circuit; locking it simply deactivates the system by breaking the connection. The only thing a thief has to do is snip the two wires leading into the lock and hook their ends together to complete the circuit's connection again.

According to the FBI, most car thieves did not even need to bother with hot-wiring the ignition because over half of all stolen cars had been left unlocked by their owners. Leaving cars unlocked has resulted in so many thefts that the *owners* of such cars can now be arrested by police on a chargeable offense in cities such as Los Angeles and Chicago.

Many of the same people who would take strong security precautions if they had $4,000 or $5,000 in the house, think nothing of leaving property worth that much unlocked in the street—and with wheels under it. Innumerable drivers are careful never to leave a full ring of keys in the ignition when they park in a public garage (having heard stories about burglars who get part-time jobs as car hikers and are later caught with hundreds of sets of house keys that they had had duplicated from drivers' key rings). The problem here is that many such house-security drivers get into the habit of never taking the key out of the ignition at all.

This applies particularly to drivers who park the car in front of the house, or in the driveway. Once they park the car in the garage, very few car owners *ever* lock up the car.

Even drivers who do lock their cars often "hide" a spare key under the hood or taped to a bumper. Experienced car thieves know all the hiding places, and they are every bit as imaginative as car owners in finding places of concealment, including such "clever" places as under the carpeting or over a sun visor. Still other drivers who lock all the doors will leave the windows open an inch or so at the top for ventilation. This is understandable, because a car, if tightly closed up, can get baking hot when left in the sun. But the practice is an open invitation to thieves. A hook made out of a wire coat hanger can be slipped into the opening to lift door-lock buttons and door handles with ease.

With so many millions of automobiles always around that are so easy to steal, a burglar may easily prefer to steal cars because there's no good reason for the average thief to steal anything that's harder to get away with.

Professional Installations

Most automotive burglar alarms are sold as do-it-yourself kits, but a notable exception is the system installed by Auto-Matic Products Company, Chicago, Illinois, which has been in the burglar alarm business for over forty years. The Auto-Matic-Alarm works on either 6- or 12-volt batteries and is controlled by a pick-proof shunt lock (with a tubular key) mounted in the fender in an armored housing. Opening any door, the hood, or the trunk activates a loud tamper-proof siren (which is illegal in California, as are all sirens used in private automobiles in that state of the Union). The installation costs $76 for a two-door car or $82 for a four-door car ($5 more for a station wagon because there's more wiring). This includes inspection service for the first year—which is $15 a year thereafter. Auto-Matic will also put in a concealed grounding switch for the ignition for $7, and a dash control switch for $7.50 to turn on the siren from inside in case of emergency.

The buyer of an Auto-Matic-Alarm also gets three "protected by" window decals. This is fine for scaring off would-be joy riders, and in fact most petty car thieves, but to a professional crook a decal only serves notice as to what he has to cope with. If he is familiar with the system noted on the decal—and as a working professional he should make it his business to be familiar with all such hardware—he will know how to disarm the system.

The one name on a window decal that is most likely to deter a thief is Babaco, the best-known name in automotive burglar alarms through long association with the trucking industry back into the days of the Depression of the early 1930s. Their patented Parker Motion Detector will sound a piercing electronic alarm siren that can be heard for a mile, as will magnetic sensors on the doors. When the alarm system is turned on at the outside lock, the ignition system is automatically immobilized, too. All wiring is concealed and protected by armored cable.

Babaco is still basically a 300-office, nationwide organization of cargo protection specialists, but they now install Parker/siren systems in passenger cars, too (mostly for salesmen, enabling them to get insurance coverage for their samples). Babaco doesn't sell the equipment, though; it's leased, with full care and maintenance contracts. Initial installation fee for a Babaco system in a passenger car might be $100 to $125, depending on complexity, plus a rental fee of $65 or so per year.

Chevrolet's Corvette is the first American car to have a built-in burglar alarm as standard equipment. Opening the doors or hood will set off a loud alarm unless the system is first deactivated with a key. General Motors also factory-installs a Theft Deterrent System in new Cadillacs as an $84 option. It has an On-Off switch in the glove compartment that the owner has fifteen seconds to shut off when he enters the car before the alarm goes off. When he parks the car and turns the system on, he has the same time delay to get out before the current sensor (see below) sets off the alarm.

Professional installations on the rest of General Motors' cars are installed by the dealer, not the factory. Most dealers use GM's Delco Burglar Alarm, a mercury switch motion detector, or a Chapman pressure-switch system with a siren and ignition cut-off costing $60 or $70. Many Cadillacs, in fact, are sold with more sophisticated systems than the regular Theft Deterrent System. One of them, which costs $160, has a motion detector activating an ignition cut-off and siren plus a Molesting Switch the occupant can use from inside the car.

Ford's factory-installed option utilizes pressure switches at each door, the hood, and the trunk to set off the alarm. The system is turned on and off by the owner with the regular door lock on the driver's side, which eliminates the "give away" lock otherwise usually set into the fender.

Do-It-Yourself Kits

The best thing about a do-it-yourself burglar alarm is that only the owner knows how it works. Even a professional thief who comes up against something he doesn't know about will usually split fast. The best-selling kits, with national distribution through auto supply stores and mail order companies, offer the least protection against a professional thief because he probably knows more about them than the car owner does. The "protected by" window decals are of most value if installed with an off-brand nobody in the area ever heard of before.

There are three basic types of burglar alarm monitors: (1) motion detectors, (2) pressure contact switches, and (3) current sensors. The sensors usually activate the car horn or a siren, but low-cost components can also be added to flash the car lights intermittently, lock the brakes, or—better yet—cut off the fuel supply from the carburetor after a few seconds so that the obviously guilty thief runs out of gas when he is a short distance down the street. Just about any burglar alarm system can be installed by the car owner himself in an hour or so for even the most complex systems, or by any garage mechanic at whatever his hourly labor rates might be in a given area.

The type of motion detector most commonly used is a simple pendulum that hangs straight down when the car remains still. The weight of anybody stepping inside the car will move that side of the car down enough to tilt the pendulum against an electric contact and blow the car horn for a startling ten-second blast. So will anybody jacking up the car to remove the wheels or tires. Motion detectors can be adjusted to be so sensitive that they will sound the alarm if anybody raises the hood or trunk lid, and even if somebody tries to pry off a hub cap. Their main disadvantage lies in the number of false alarms. The pendulum sensor, mounted flat to the fire wall to detect any side-to-side motion (or against the radiator in cars with air conditioning where there's no room on the fire wall) can be set off sometimes by mere gusts of wind or even by fast-passing traffic if the contacts are too close. Even cheap pendulum-type motion detectors have adjustment knobs to control the sensitivity, depending on the weight of the automobile.

Dependable mercury contact switches, which still cost only about $2.75 apiece in hardware stores, can be mounted in pairs to detect even more types of movement than the side-to-side pendulum. The tilt angle on a mercury contact switch is also fully adjustable, depending on the car size, to prevent false alarms—unless somebody tries to sit on one of the fenders.

Many motion detector kits, complete with wiring and an On-Off switch for the dash, are sold for less than five dollars by auto supply companies such as J. C. Whitney, Chicago. The car owner should have a horn relay to boost the horn for a louder sound (most Fords do not have one), but even a heavy-duty relay is only about two dollars, built to withstand the heat generated by continuous operation so it won't short out.

Whitney also sells a $12.88 kit for a two-way system that incorporates a motion detector with plunger-type contact switches, which work like the button-switch on a refrigerator that turns on the interior light when the door is opened. A separate On-Off switch allows the use of either system independent of the other; the wiring furnished has its own in-line fuse; the activated alarm can be shut off only with a key in a special lock; and the kit even includes window decals (don't use them if you think a professional might be after your car).

About as deluxe as Whitney gets is the $24.95 kit using an all-switch system, one for each of the four doors and one each for the hood and trunk. Any of the six contact switches will activate a police-type siren, and a pick-proof lock mounted flush on the outside of the car body sets or silences the system with a turn of the tubular key. One of the switches can also be wired to the emergency brake to protect convertibles with the top down.

Basically the same six-switch system, but having an ordinary lock with plain keys, is $19.88 with a smaller siren. A five-switch system, with a bell instead of a siren, is $14.95. The car owner who wants to rig up a one-of-a-kind system of his own can buy plunger-type contact switches for about sixty-nine cents apiece. He can get a siren from $30 or so for a "yelping type" siren activated electronically (an ordinary siren can be silenced by jamming the impeller) down to an $8.98 "mini" siren about the size of a pack of cigarettes, which is easy to hide. Separate alarms are recommended for cars that have the horns mounted so low that a thief can reach the wires from underneath; Californians, who can't legally use sirens, can always buy a spare horn, often for as little as a dollar or two from a wrecking yard.

Montgomery Ward sells a $29.98 six-switch kit that incorporates an added security feature, the ignition cut-off. This prevents a thief from stealing the car even if he does have enough guts to drive away with the siren wailing. Without the ignition cut-off, it's $19.98, and the kit for hooking up the horn (no siren) costs $7.98.

Chicago's A. W. Fruh & Company, one of the best-known specialists in security equipment, has a $39.95 list price for its standard siren-equipped six-switch burglar alarm, but it includes a genuine Ace pick-proof lock, for which the tubular keys cannot even be duplicated by an ordinary locksmith. The comparable Ademco six-switch kit installs for anything from $50 to $100, depending on which installer does the selling to the car owner.

One type of contact switch, not yet available in kits for car owners, is the mat switch (see Chapter 6). This can be particularly effective because most thieves do not think of mat switches in terms of automotive security and will not know how they have activated any alarm they have set off. Placed under the carpeting in front of the driver's seat, a mat switch can be wired into any auto alarm system to sound the horn or siren when the thief puts his feet on the floor, even if he has gotten by the perimeter protection by, say, ripping through a convertible top.

Among the most sophisticated burglar alarms are the systems using electric current sensors, touted by some as

Most burglar alarms for automobiles are do-it-yourself kits, but the ones utilizing sirens are illegal to use in some states, such as California.

computer-controlled alarms that protect the car by electronics. Ademco's customers (the salesmen/installers) get as much as $75 to $150 for the system—but it's a good one.

The "wireless" alarms sound the siren whenever the electrical system of the car draws current. This will happen when interior lights in the car go on, such as dome lights or courtesy lights that go on when a door is opened, when the trunk lid is opened, or even if the glove compartment is opened. Stepping on the brake pedal, which lights up the taillights, will also activate the alarm, and using the engine starter, of course, draws the most current of all from the battery and will set off the siren even if a malfunctioning switch or a burned-out dome light has allowed the thief to get into the car unnoticed.

Ademco's wireless alarm system will continue to sound

the alarm even if the activating door is immediately closed, and it can be turned off only with the owner's key. The sensitivity to the current drain on the system is preset at the factory so that low drain devices such as electric clocks will not set off false alarms. The installer can adjust the sensitivity even further by controlling how close the alarm power line is wired between the battery and the voltage regulator.

This wireless system is extremely popular with installers because it requires almost no installation wiring, inasmuch as it makes use of existing wires in the car's electrical system. The only thing missing: security for the hood. A standard plunger-type contact switch up there and wired into the system, though, is all that's needed. Em-D-Kay's "howler" differs from Ademco's wireless system principally in the method of distribution, with the Brooklyn company selling their package in a do-it-yourself kit.

James Electronics, Chicago, Illinois, puts up a comparable current-sensor system in kit form, selling at $39.95 to the do-it-yourself car owner. It even has an advantage over the Ademco model in that it temporarily disables the ignition, without damaging the coil. However, it has no siren, and depends on the automobile horn for its audible alarm.

The solid-state James Kar-Safe doesn't have an exterior lock to turn the system on and off, either. Instead, it is programmed to wait several seconds, as on the Cadillac Theft Deterrent System, before the horn starts blowing after the owner leaves or enters, using a simple hidden manually-operated On-Off switch behind the dash. When leaving the car, the owner simply opens the door and flips the switch to the On position. The time-delay unit does not become armed until after the door is closed. When reentering the car, the owner has seven seconds to switch the system off. There are no keys to carry or expensive installation costs for putting in a lock.

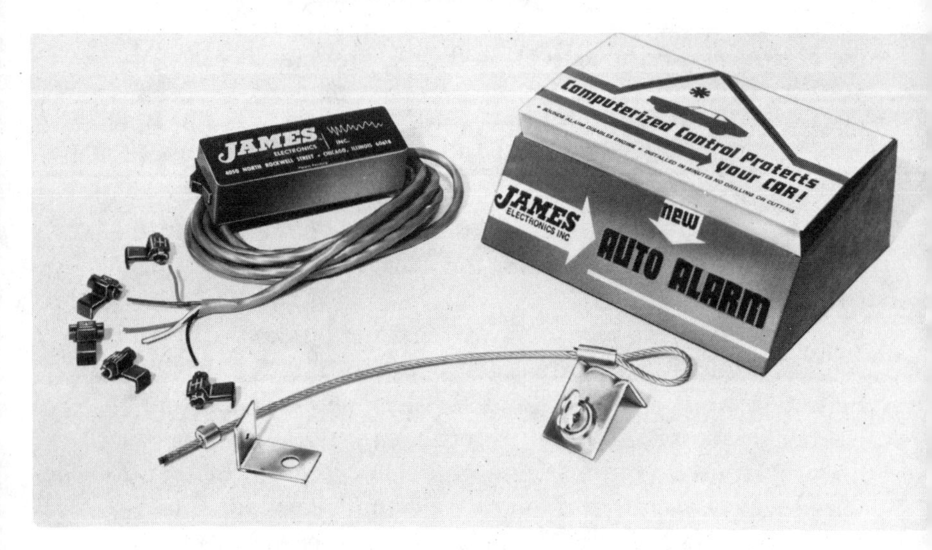

The James Auto Alarm not only sets off the car horn but also disables the ignition so that the car cannot be started.

Because there are no visible signs of an alarm, a time-delay switch also eliminates one of the big disadvantages of most systems using an exterior lock to arm or deactivate the burglar alarm—the exterior lock is not only a dead giveaway that the car is alarm-equipped, but when mounted in the fender it is also vulnerable to a knowledgeable thief's reaching up and cutting the wires behind it.

Another type of burglar alarm that can be activated with any type of sensor is a high-frequency radio system. The owner carries a small receiver, shirt pocket size, which starts beeping madly if anybody starts meddling with the car if he is anywhere within two blocks of the car. Components for the transmitter-and-receiver system are available from electronics supply companies, such as the Radio Shack chain throughout the country, for about $100. A radio-alert system is for car owners who do not have much faith in the average passerby's willingness to interfere with anybody who sets off a car's horn or even a siren. In such cases, some people figure, the most sophisticated siren or horn alarm is no more effective than the fifty-cent practical joke sold in novelty stores that starts whistling and billowing smoke from under the hood when the car engine is started. Perhaps it is not even as much protection as the simple S-shaped steel shaft that locks the steering wheel in place by rigidly connecting it to the clutch or brake pedal, and costs only about eight dollars from Western Auto or Sears Roebuck, including two keys for the shaft lock.

Component Locks

An increasing number of people who leave their cars parked outside overnight wake up in the morning to find that they have been stripped of hub caps, radios, stereo tape players (in particular), batteries, tires, special wheels in their entirety, and even the bucket seats, with the trunk looted besides.

Most car looters do not have the know-how to pull a door lock, and using the classic coat hanger hook through the top of a window is the extent of their expertise. One way to foil them is to replace the original-equipment door-lock buttons that have mushroom-shaped heads or the equivalent to make pulling them up easier, as thieves well know. Smooth-sided replacements, which do not have a surface for a hook to get a grip on, cost only about eighty-nine cents a pair. They are admittedly tough to install because the doors have to be taken apart. Buying an eight-dollar lock mount is still a good idea for anybody with a stereo, radio, or citizen's band radio set, which makes it removable and portable.

A simple-to-install key-operated battery lock (Martin Auto Parts, Chicago, Illinois) replaces the hold-down nut to prevent theft of this valuable-anywhere auto part that is so frequently stolen from unattended automobiles. Similar locks, from Universal Security Instruments, Inc., Balti-

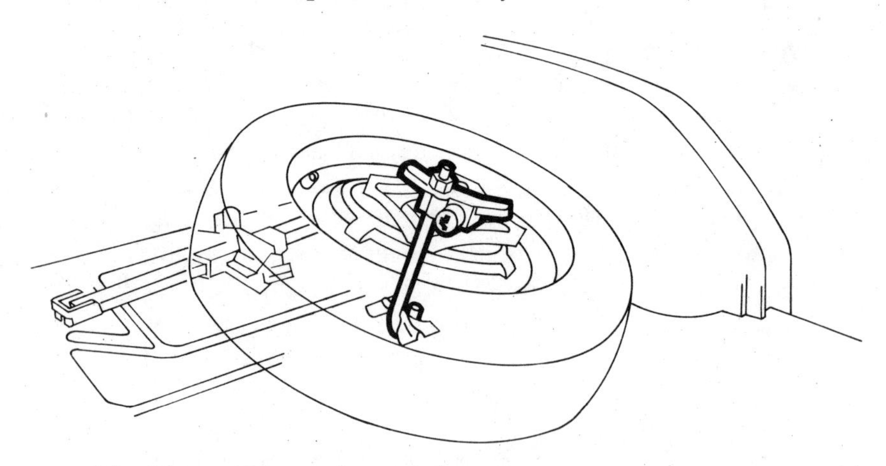

Spare tires, which are often new,are a prime target of thieves, and a lock for the spare tire can be a worthwhile investment.

more, Maryland, can prevent wheels from being stolen by replacing one of the lugs in each wheel with a key-operated replacement.

A better system for protecting batteries and other under-hood equipment, though, is to lock the entire hood. For cars with hoods that can be opened normally from outside, Whitney has a cable-operated accessory that locks and unlocks the hood from inside for $3.49, complete. Wards and Sears (among many others) have key-operated hood locks that can direct-lock the hood from outside for less than five dollars.

Wards and Whitney both sell electric hood locks for eight and nine dollars. This type of lock releases the hood with an inside push-button that operates only when the ignition is turned on. Whitney's solenoid model has a second push-button switch, hidden in the grille, which also must be pushed before the hood can be opened with the regular latch release.

Whitney also has another hood lock that is unique in that it shorts out the ignition system for good measure, if the

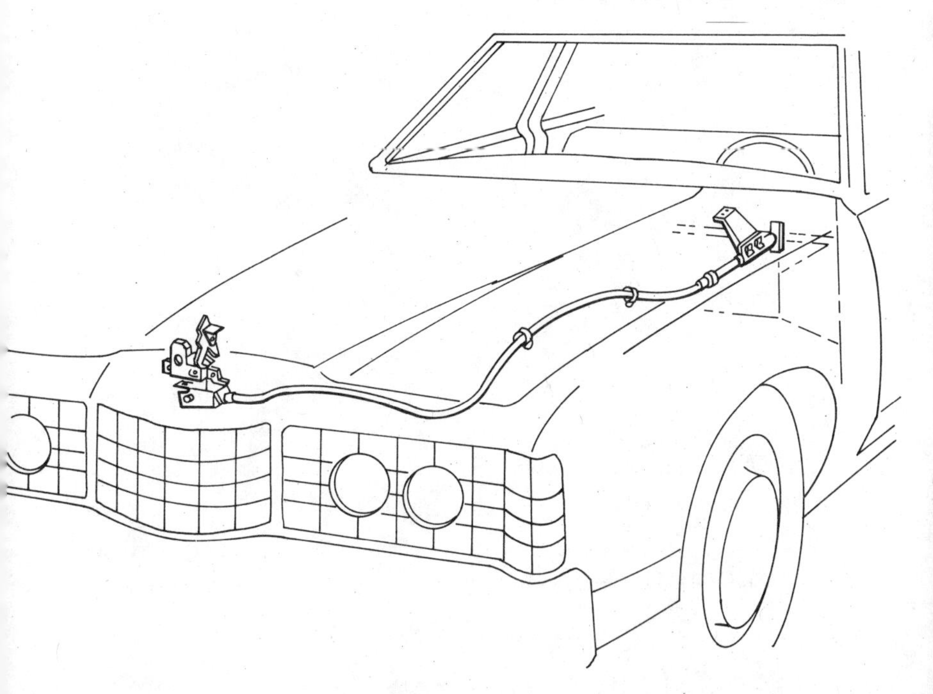

One of Ford's security options is an interior hood release to keep thieves from getting at the battery and other under-hood components that are worth stealing.

hood is tampered with. The $12.95 system locks with a push button, but it can be unlocked only with a key. Even if a cutting torch is used on the locking mechanism, and a chain cutter is used to cut the cable, the hood still will not release. The ignition stays shorted out even if the wires are cut, and it cannot be hot-wired.

Whitney has still another ignition-activated "undercover" lock, for $9.98, that can be used to keep either the hood or the trunk locked. Normal closing of either automatically locks the compartment until the owner turns the ignition key. A separate $9.95 electric release, designed specifically so that the trunk lid can be opened from inside

Lock-equipped wheel lugs can prevent car-strippers from stealing the wheels if they jack up the car.

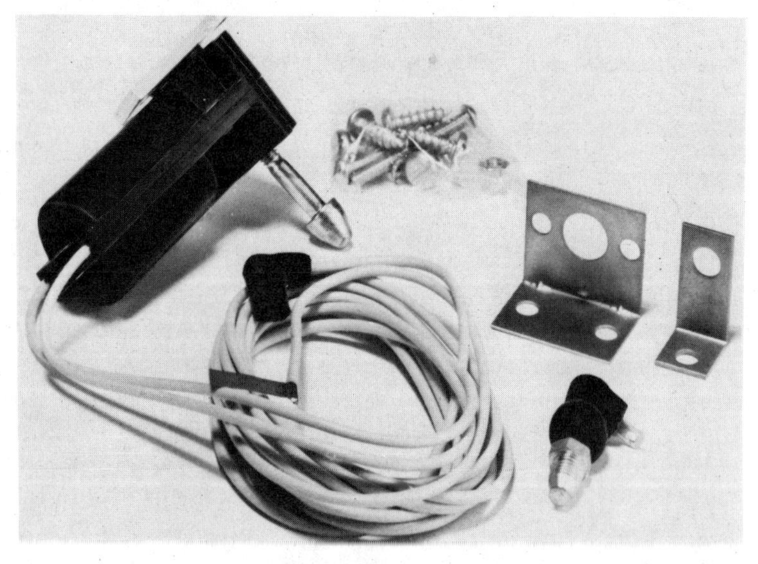

Whitney's hood lock kit sets with a push button but can be unlocked only with a key. It will short out the ignition if tampered with.

the car with a push button or from outside with the regular key, fits most GM cars but not all Ford products. The $7.75 all-electric trunk kit, which does away with the trunk keys altogether, fits even fewer cars.

During the worst of the energy crisis (so far), thousands of car owners installed key-locking gas caps to foil siphoners equipped with a length of surgical tubing. These accessory caps are now a glut on the market, and discount houses sell them for as little as $1.25 apiece, with models even for foreign cars seldom costing more than $5 for pretty good two-key locks.

Motorcycles

The motorcycle industry builds about a million bikes a year for the American market—where nearly half a million a year are stolen. The customized choppers and cafe racers are relatively safe from professional thieves because they are too individually recognizable. But the standard Honda, Suzuki, Kawasaki, Yamaha, Triumph, Norton, Moto-Guzzi, Benelli, BMW, Husqvarna, Bultaco, Montesa, or Harley finds ready buyers in the stolen bike market, which includes plenty of minibikes, too.

The standard ignition lock on most motorcycles is little more than a convenience item; crossing the wires to complete the circuit without a key is so simple that it can happen accidentally. Even if it's locked, the bike can be rolled away all too easily.

A standard operation for professional thieves is to push the bike up a portable ramp into a waiting truck. If they use a more unobtrusive pickup truck, they throw a tarp over the first stolen bike while heisting the next one, sometimes collecting three or four per trip.

The rider who wants to keep an unattended bike from being stolen has only one dependable course of action: chain it down, to something immovable. Most hardware stores carry tempered, case-hardened chain with links

All security chains should be long enough to allow wrapping around a large stationary object such as a lamp post.

heavy enough to make breaking the chain difficult. A 3/8-inch chain costs up to two dollars per foot, but it's a good investment.

Any chain can be cut with a big enough bolt cutter. Do not be dismayed when you see the hardware clerk cut your length of chain from a large roll with an ordinary bolt cutter—the ordinary joyrider does not carry such equipment. And he is the only one you can stop from stealing your motorcycle. All you can do about a professional thief is slow him down a little.

The weakest link in an anti-theft chain is often the padlock (see Chapter 2 for details on types and styles).

Suffice it to say here that the shackle on a padlock should be made of case-hardened steel, and the cylinder should have a minimum of five pin tumblers. A heel-and-toe shackle, with locking notches on both sides, provides double security. Good locks start at about three dollars, but the average cost of padlocks used on motorcycles is closer to ten dollars.

Bike riders, too, worry about their fuel supplies. Whitney has a key-locking gas cap with a fuel gauge in it (not standard equipment on motorcycles) for $9.99. They also have a $4.98 gas cap with a fake key slot, which may give a thief more trouble than a real lock.

Bicycles

At one time, stealing bicycles was mostly a lark for kids. Most bikes are still stolen because their owners carelessly leave them unlocked. But now there are professionals in the bicycle field, too, going after the popular imports and 10-speeds; some of them cost $400 or more, and are well worth stealing.

Woolworth's sells bicycle security locks that are not far from being fraudulent, with Kresge's and the Ben Franklin chain being only a little less guilty in selling locks that can be opened with a good swift kick. The fast-selling long-shackled lock for the front fork will keep the front wheel from turning but will not prevent anybody from carrying the bike away. The chains sold in some sets are often no heavier than the chaining used to hang porch swings, and some of the combination locks on chain sets are even made of soft white metal.

Chains and lock shackles should both be 9/32 of an inch thick for good bicycle security. The chain itself should be in a plastic casing to protect the finish on the frame. The chain or cable should be at least three feet long so that it can be wrapped around big stationary objects. Master Lock Company makes an interesting self-coiling cable that stretches out to a full six-foot length, but which snaps back

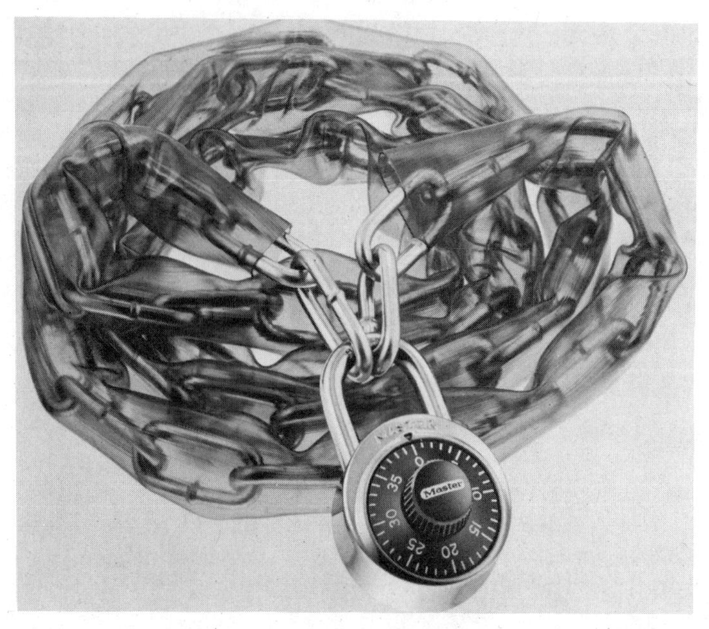

Lock-up chains should be enclosed in protective plastic casings to prevent the scratching of the bike's finish.

into a compact coil when released.

The best bicycle locks are sold by the Turin Bicycle Shops, now operating in most major cities. Their Kryptonite lock, with an amazingly pick-proof cylinder, is virtually theft-proof because the lock itself is nearly surrounded by heavy steel jaws. Instead of utilizing a chain, the Kryptonite has its own U-shaped closure made of heavy strip steel no bolt cutter can work on, big enough to clamp the bike frame to a parking meter. But a Kryptonite lock retails for $17.

Even the best conventional lock is practically useless if used with a chain that is less strong than the lock itself. A chain that heavy could weigh as much as forty pounds—and few people are willing to pedal that much extra weight around. The result is that expensive bikes locked up with

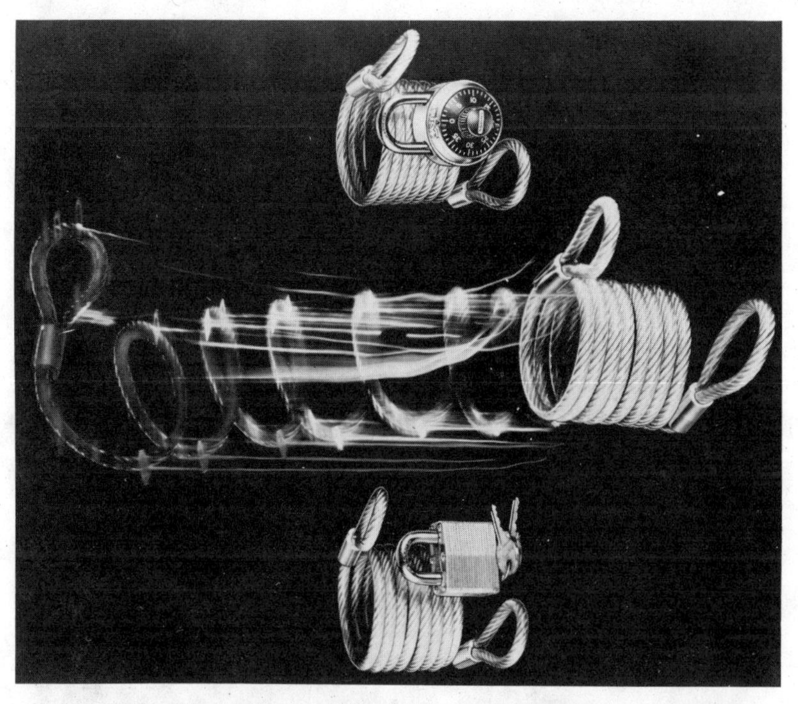

Unique self-coiling cable made by Master Lock stretches out to a six-foot length but snaps back to compact coil when released.

good locks are stolen left and right by "chain gangs," leaving the lock untouched.

A bicycle "chain gang" consists of a tool man, equipped with a bolt cutter carried in a paper bag, walking down the street with his riders trailing him about a block behind. A strong man with a bolt cutter can cut a link in all but the strongest chains in less than a second. He is safely away with the incriminating tool, guilty of nothing very serious, by the time his rider reaches the now-unchained bicycle.

If the rider doesn't see anybody watching him, he hops on and rides off, leaving the cut chain on the ground behind him. If he is caught in the process, he can claim that he was

Kryptonite bicycle lock utilizes horseshoe-shaped clamp instead of a chain, and its steel jaws are impervious to attack by the bolt cutters used by most professional bicycle theives.

only doing a little joyriding on a bike he found unchained. "I was gonna bring it back, honest." If he is caught more than once or twice, the tool man replaces him with a new and cleaner rider.

About the safest thing the owner of an expensive bike can do is take steps for recovery if it's stolen (see Chapter 15). The best protection a bicycle rider has, by long odds, is to ride an old coaster brake model with mismatched wheels and a rusty frame. Nobody will want it. The owner of such a contraption will have one of the most dependable laws in the world working for him—the law of supply and demand.

Recreational Vehicle Protection

To the layman, a "camper" is any vehicle used by vacationers to sleep in. In the trade, a camper is technically a slide-in unit mounted on the back of a pickup truck, and other recreational vehicles include trailers, van conversions, and motor homes. But as far as this book is concerned, they are all treated as campers. They are on the roads by the millions, and they get broken into and/or stolen by the hundreds of thousands for four main reasons: (1) they are often loaded with cameras, radios, liquor, and other goods attractive to thieves; (2) they are often left parked in isolated areas or in campgrounds where nobody knows the owners and thieves can work undisturbed; (3) the victims are usually from out of town and do not stay around to urge investigations or press charges; and (4) the basic security systems in most recreational vehicles are almost nonexistent.

The recreational vehicle industry is growing at a phenomenal rate and is full of fast-buck operators trying to get rich quick. Many manufacturers use the cheapest components they can get, including door locks of the sleaziest kind, doors themselves so flimsy they can be cut open with an ordinary can opener, and sometimes windows without any locks at all. Some thieves have even been known to cut their way in through the thin plywood floor or even through the aluminum walls with tin snips.

There wouldn't be much point in putting a big padlock on the door of a tent (and indeed, most of the low-profile folding "camping trailers" do have canvas walls). But there's plenty of point in putting a good loud burglar alarm on a camper. The people who stop in campgrounds may not know each other, but most of them do have more of a sense of camaraderie than some of the people in big cities, and they *will* investigate if they hear a burglar alarm sounding. In the case of recreational vehicle protection, the window decals do more good than harm.

Whitney's recreational vehicle alarm system incorporates a motion detector with door and window switches, in a kit identical in basic operation to their two-way automobile alarm (see above). The motion detector protects against towaway attempts and the switches guard against forced entry. The difference in the $29.88 recreational vehicle model is that it has a loud 85-decibel police-type siren. The same basic kit is also available for $5 less with a clanging bell instead of the siren.

Wards has a recreational vehicle burglar alarm that is entirely self-contained in a 3 x 6½ x 4½-inch metal cabinet designed to be mounted on the inside of the door. The fully-transistorized electronic unit, which costs $29.49, requires no wiring, contains its own power source, and has its own shrill yowler in the tamper-proof casing.

Bigger recreational vehicles can also use some of the burglar alarm systems designed principally for use in the primary residence (see Chapter 6). A family owning a recreational vehicle should also take security precautions when it is not in use between vacation trips and sitting neglected. Many a camper owner has gone to get it ready for a vacation only to find that it has been sacked.

Planes and Boats

Anybody who can afford to own private aircraft or a good-sized boat can afford to buy good security equipment —and he needs it. Planes and boats are designed for the pleasure of their owners, with security getting very minor if any consideration. The cabin windows on most sailing craft are too small for an intruder to get into, but the hatch is most often secured with the meagerest of locks. The bigger windows on power boats are most usually sliding windows with only a simple catch—much easier to force with just a nail file than even a double-hung residential sash window.

In the interests of light weight—seemingly the *only* consideration in the construction of many models—the skin on the average private plane is so vulnerable to attack by a thief that it can be readily cut with nothing more than a sharp hunting knife. Ignition locks are usually nothing more than convenience hardware.

One of the most surprising things about planes and boats for a security-conscious observer is that most of them are already equipped with audible emergency alarms—but seldom attached to security sensors. Most sizable boats have foghorns and many small planes have warning howlers to indicate carelessly used landing gear or other malfunctions. Any of these are easily wired into sensors that can detect unauthorized entry or movement.

Burglar alarm systems designed specifically to protect expensive boats and private planes are usually expensive—perhaps on the theory that an owner can afford the expense. Fruh, one of the largest in the field of security equipment, gets $170 for its self-contained alarm, or $185 with a shunt lock in the 6½-inch square unit.

It's a good one, though. Underwriters' Laboratories has passed it in tests at 30° F. below zero, 160° F. heat, and for abnormal dust, humidity, vibration, and sleet or rain conditions. Its alkaline battery has a one-year life in all weather, with a 35-ampere hour capacity. The tamper-proof case is made of 16-gauge stainless steel with its own pry-off contact. It can monitor any number of door, window, fuselage, or hull contacts.

The leader in the field of marine security is Aqualarm, Inc., Gardena, California. The basic $49.50 anti-theft and anti-vandalism kit consists of two pairs of magnetic switches (see Chapter 6), an exterior key-lock switch to arm or disarm the system, and a tamper-proof control box con-

Aqualarm anti-intrusion kit is most usually hooked into a total security system warning against everything from low oil pressure to hull flooding.

taining a battery-operated bell-type burglar alarm. It is most usually wired into a total-security system with a fire and bilge protector ($59.50) that operates on dry or ship batteries.

The Aqualarm system covers hull flooding, engine overheating, and low oil pressure, with optional accessories (most under ten dollars) that can detect anything from low water pressure to overheating with remote sensors. Aqualarm has been so successful in the field of marine protection that the company is now starting to distribute nationally a comparable $69.50 kit for basic residential protection in homes, just as Master Lock Company is doing (see Chapter 6). A boat owner has a sizable investment to protect, and installation of a burglar alarm can even reduce insurance premiums in what most underwriters consider a high-risk field, usually charging high insurance rates accordingly.

14

Property Recovery

"What's the point in stealing something if I can't sell it? The heat even knew who the stuff belonged to the last time they caught me with a stash"—Inmate, County Jail, Los Angeles, California.

One Zenith television set looks like any other one of the same model. Police recover a tremendous amount of stolen property that the victims cannot identify—and when that happens, the stolen property is often returned to the thief the cops took it away from. Detectives in Brooklyn recently nabbed a burglar in the act, and when they checked out his apartment they found dozens of cameras, radios, electrical appliances, and wristwatches. But nobody could identify any of the property with enough positive identification. The burglar's attorney took it for his fee.

Engraving Pen Program

A nationwide Property Identification Program is now in full swing, sponsored by the Federal government. Police departments in all major cities and many minor ones are supplied with electric engraving pens that can permanently mark possessions to discourage theft. The program is publicized through civic organizations and local newspapers, urging all citizens to stop in at the police station and borrow one of the pens, free of charge, for up to a week. In Omaha, enthusiasm grew so high that the Boy Scouts went around in a door-to-door campaign to promote the idea.

The only thing anybody needs to do to borrow one of the pens is show identification proving that he is a citizen of the town. He even gets a window decal, serving notice that he is participating in the Property Identification Program and that valuables in the house are indelibly marked accordingly.

The engraving pen is as easy to use as an ordinary pencil. It can be used to write or draw on almost any material: wood, metal, plastic, or glass. The ultra-hard tungsten carbide tip vibrates at 7200 SPM, with a variable stroke adjustment for heavy or fine line engraving. A UL-listed 115V AC model, in a high-impact housing, is available from hardware stores and mail order companies at $6.95 for people who live in areas where the police departments are not participating in the federal program.

The biggest lack of participation is not on the part of "too busy" police departments, though, but on the part of the public. Stories go around at police conventions that some cities have reduced burglaries by as much as 75 percent when they got total public participation in the engraving pen program, but this is fiction put out by the pen manufacturers and the bureaucrats running the program. Most police officials privately admit that 3 to 5 percent participation—a long, long way from "total"—represents a *successful* campaign. In some towns, none of the pens have ever been used at all—and the program is almost two years old.

As for the program being a burglary deterrent, no chief of detectives can be found who will say that it has reduced crime in any traceable way. Most burglars work fast, and they do not take time to examine what they're stealing to see if it has been specially marked. If they do find later that the stolen property has indeed been marked, they can always blot out the marking with the same kind of engraving pen the owner originally used (the engraving pens are available on loan to *any* citizen, and nobody asks what he's going to do with it).

Identification Numbers

Most property owners who do use the pens use their driver's license number. When any property is found with that number on it, it can be traced easily through the state's secretary of state to find the owner. Another easy number to trace would be the automobile's state license plate number; most sizable police departments have directories listing every current license plate issued in the state, with the name and address of each car owner.

Unless the car owner requests the same number every year, however (and pays an extra fee for the privilege), the numbers on the license plates change every year. Even using the number on the driver's license has its disadvantages (not counting the many people who don't have one), because the population is now so fluid that almost 20 percent of the people move every year, and often to another state.

A social security number can be used to identify stolen property, but it is useless to the police in tracing an owner because the Social Security Office will not release any information for such purposes. All in all, the householder who marks his property with his name and postal zip code, as well as with the date, has a system that's as infallible as any.

In some cases, the property cannot be marked at all, as in the case of rings and other small pieces of jewelry. It should be photographed, instead, for identification purposes if recovered after having been stolen.

The federally sponsored Property Identification Program, when operated with the full cooperation of local police departments, can be of great value if it is accepted within its pragmatic limitations. But when considered as a cure-all like an oversold burglar alarm or any other single security precaution, it can be misrepresented for what it cannot do by itself.

One thing that makes the engraving pen program look like something of a federal boondoggle is the fact that most valuable equipment already has individual serial numbers on it, as on everything from cameras to typewriters. Serial numbers can indeed be altered or taken off, but so can any other kind of marking. What the property owner has to keep firmly in mind, along with positive identification, is registration. A marking system without it is relatively meaningless.

Bicycle Registration

One of the best examples of the advantages of registration is the Los Angeles Police Department's record with stolen bicycles. Bicycle owners in L.A. are urged to come in and register the bike at the police station. As in most cities that issue bicycle licenses, the registration is not mandatory; it's just a good idea.

The owner gets a numbered tag for his bike for the small registration fee, and he has almost as much protection as the owner of a licensed automobile or motorcycle. Over 75 percent of all registered bicycles stolen in Los Angeles are returned to their rightful owners—while only 2½ percent of unregistered bikes are recovered. Many thousands of abandoned bicycles—presumably stolen by joyriders—are sold by the police at public auction every year all over the country.

Stolen Cars

The recovery of stolen cars is complicated by the ease with which so many are sold to innocent used-car buyers. But the innocent buyer of a hot car can get into a lot of trouble, and will also, of course, lose his investment if the car is reclaimed by the driver it was stolen from.

The buyer of a used car should beware of the bargain-priced cream puff if the seller is not a reputable dealer. A tip from a casual acquaintance, or a notice tacked to a tree or in the window of a car parked on the street, may result in the buyer's finding out when he applies for a license that he will have neither the car nor the money he paid for it.

A car thief getting rid of a stolen vehicle "on the street" will always pitch for a quick sale. He will also give vague replies when asked specific questions, or give answers that don't jibe with what the prospect learns later on.

If the seller doesn't work for a used-car agency with an established reputation, the buyer should always ask for a home or business address that can be readily checked. Ask flat out where the guy bought the car, and where he's been having it serviced—and check it out.

License plates should match the car. Dirty plates on a shiny new car, or new bolts on old plates, are indications of a license plate switch. Duplicate keys are also a suspicious sign, as a car thief is not very likely to have the original set. A close look at locks can show signs of tampering or replacement if the original lock has been pulled. *All* keys should work properly in the proper locks. Tool marks or chipped glass on any of the windows or vents are also signs of break-in, as is any replaced glass (with the trademark not matching the original).

The engine number should be checked for evidence of grinding, restamping, or alteration. Documents should be checked for altered figures such as an 8 that might have been a 3. If the date of previous sale on the certificate of title or bill of sale is recent, the buyer should find out why the rush in selling the car. The figures in all documents should be cross-checked against engine numbers and license plate numbers. If there are any signs of forgery or alteration, the seller's signature should be compared with the signature on his driver's license and license identification card.

If the prospective buyer of a used automobile thinks that somebody's trying to unload a hot car on him, there's only one thing to do: call the cops. The recovery of stolen property should be everybody's business, whether it's his property that's being recovered or not.

There's also plenty of "hot" property being sold that is hot in name only. Every big city has a lot of peddlers sneaking around in saloons or hanging around factory gates who pretend to be fences for wares that they actually buy at wholesale. The suckers who buy this usually shoddy stuff think they are getting "a price that's a steal" when, in fact, they are *not* getting it for what they've been told is half of what it's worth; more often they pay twice what the same merchandise would cost in a store. The more ingenious of these con men even file the serial numbers off their "stolen" wristwatches to make the story more plausible.

Personal Inventory

Serial numbers on a householder's property, or any other identifying marks, are meaningful only if he keeps a record of them. The homeowner should make a detailed list of all valuable property, roomful by roomful. And in today's social climate, *everything* in the house should be catalogued, especially now that burglars sometimes drive up to a well-cased house with a moving van and move out the entire contents, right down to the bare floor.

Hopefully, the homeowner will never need to use such a list. But it is also good protection in case of a fire loss, as well as losses to burglars. Many insurance companies won't pay off unless there *is* such a list.

The following pages may suggest how much more property the average householder owns than he might realize. The pages can be filled in right here, according to the different kinds of possessions the homeowner has, but they should not be left in the book (which is a grab item for burglars who are interested in keeping up with what's going on in their trade). Cut out the filled-in pages and keep them in a safety deposit box along with the receipts or delivery tickets.

The householder who does so has a much better chance of recovering his property.

Article	Brand name	Model numbers
LIVING ROOM		
Television set		
Hi-fi or Stereo		
Record albums		
Piano		
Air conditioner		
Clocks		
Fireplace tools		
Paintings		
Davenport		
Chairs		
Tables		
Mirrors		
Lamps		
Rugs		
Carpeting		
Drapes		

Supplier	Cost	Serial numbers or identifying mark	Date of purchase

Article	Brand name	Model numbers
DEN		
Typewriter		
Tape recorder		
Cameras		
Projectors		
Filing cabinet		
Calculator		
World globe		
Books		
Television set		
Liquor cabinet		
Air conditioner		
Golf clubs		
Bowling bag		
Firearms		
Trophies		
Clock		
Desk		
Chairs		
Lamps		
Carpeting		
Drapes		

Supplier	Cost	Serial numbers or identifying mark	Date of purchase

Article	Brand name	Model numbers
DINING ROOM		
Silverware		
Tea set		
China		
Stemware		
Breakfront		
Buffet		
Table and chairs		
Light fixtures		
Air conditioner		
Chafing dish		
Mirror		
Carpeting		
Drapes		

Supplier	Cost	Serial numbers or identifying mark	Date of purchase

Article	Brand name	Model numbers
KITCHEN		
Toaster		
Waffle iron		
Broiler		
Blender		
Food mixer		
Can opener		
Electric knife		
Coffeemaker		
Radio		
Clock		
Air conditioner		
Dishwasher		
Refrigerator		
Stove		
Freezer		
Cutlery		
Table and chairs		
Fire extinguisher		
Vacuum cleaner		

Supplier	Cost	Serial numbers or identifying mark	Date of purchase

Article	Brand name	Model numbers
MASTER BEDROOM		
Jewelry		
Watches		
Furs		
Clothing		
Gloves		
Wallets		
Boots		
Binoculars		
Luggage		
Pistol		
Clock radio		
Television set		
Sun lamp		
Air conditioner		
Humidifier		
Sewing machine		
Fire extinguisher		
Bed		
Dresser		
Vanity		
Tables		
Lamps		
Chairs		
Bedding		
Paintings		
Carpeting		
Drapes		

Supplier	Cost	Serial numbers or identifying mark	Date of purchase

Article	Brand name	Model numbers
CHILDREN'S ROOM		
Record player		
Albums		
Musical instruments		
Sports equipment		
Air conditioner		
Clock-radio		
Bed		
Blankets		
Desk		
Chairs		
Dresser		
Lamps		
Carpeting		
Drapes		

Supplier	Cost	Serial numbers or identifying mark	Date of purchase

Article	Brand name	Model numbers
BATHROOM		
Drugs		
Electric razors		
Electric toothbrushes		
Hair dryer		
Bath whirlpool		
Scale		
Vaporizer		
Rugs		

Supplier	Cost	Serial numbers or identifying mark	Date of purchase

Article	Brand name	Model numbers
BASEMENT		
Bench saw		
Power tools		
Soldering iron		
Motors		
Pumps		
Welding equipment		
Compressor		
Vise		
Hand tools		
Tool chest		
Cord sets		
Paint brushes		
Ladders		
Work bench		
Stools		
Lighting		
Fans		
Heaters		
Clock		
Radio		
Fire extinguisher		
Washing machine		
Dryer		
Ironing board		
Laundry iron		
Pool table		

Supplier	Cost	Serial numbers or identifying mark	Date of purchase

Article	Brand name	Model numbers
GARAGE		
Lawn mower		
Snow blower		
Hedge trimmer		
Chain saw		
Garden tools		
Ladders		
Battery charger		
Tune-up kit		
Garage jack		
Lube guns		
Oil supply		
Tire chains		
Bicycles		
Snowmobile		
Outboard motor		
Fishing gear		
Tent		
Coleman lantern		
Sleeping bags		
Picnic gear		
Lighting		
Fire extinguisher		
Automobile		

Supplier	Cost	Serial numbers or identifying mark	Date of purchase

Index

G

H

I

T